SEXUAL SEGREGATION IN UNGULATES

Wildlife Management and Conservation
Paul R. Krausman, Series Editor

Sexual Segregation in Ungulates

Ecology, Behavior, and Conservation

R. TERRY BOWYER

Published in Association with *THE WILDLIFE SOCIETY*

JOHNS HOPKINS UNIVERSITY PRESS | BALTIMORE

© 2022 Johns Hopkins University Press
All rights reserved. Published 2022
Printed in the United States of America on acid-free paper
9   8   7   6   5   4   3   2   1

Johns Hopkins University Press
2715 North Charles Street
Baltimore, Maryland 21218
www.press.jhu.edu

Library of Congress Cataloging-in-Publication Data

Names: Bowyer, R. Terry, author.
Title: Sexual segregation in ungulates : ecology, behavior,
    and conservation / R. Terry Bowyer.
Description: Baltimore : Johns Hopkins University Press,
    2022. | Series: Wildlife management and conservation |
    Includes bibliographical references and index.
Identifiers: LCCN 2021063011 | ISBN 9781421445069
    (hardcover) | ISBN 9781421445076 (ebook)
Subjects: LCSH: Ungulates—Behavior. | Spatial behavior
    in animals. | Sex discrimination.
Classification: LCC QL737.U4 B69 2022 | DDC 599.6—
    dc23/eng/20220120
LC record available at https://lccn.loc.gov/2021063011

A catalog record for this book is available from the British
Library.

*Special discounts are available for bulk purchases of this
book. For more information, please contact Special Sales
at specialsales@jh.edu.*

To my wife, Karolyn
Nothing would be possible or enjoyable without her love and support

Contents

Preface

For as long as I can remember, I have been fascinated with large mammals. I suspect this attraction originated with my dad taking me out of school so I could accompany our extended family on hunting trips for mule deer in California's Sierra Nevada. At that time, I was far too young to carry a rifle. Nevertheless, I read everything I could about hunting, and when mature enough to hike through rugged terrain and shoot straight, I joined my dad hunting mule deer and quail in the mountains outside of Ojai in southern California. Early on, my enjoyment of hunting helped me decide that I wanted to work with large mammals. I was studying for my BS degree in wildlife management, however, before I recognized that being a hunter, although teaching me field craft, was only a small part of what was required to be a scientist. Nevertheless, without that early background, I might never have become a field biologist.

I initially was not interested in sexual segregation; I never recognized that this topic was a major component in the life histories of many large mammals (incidentally, hunters are not the only ones who lack this appreciation). I had spent much of my time in the field during autumn hunting seasons, which coincided with rut, when the sexes were mostly aggregated. Later on, I began to enjoy watching animals, especially ungulates, without disturbing them, and I was becoming increasingly more curious about their behaviors.

I was fortunate that my master's research at Humboldt State University was supervised by David W Kitchen, an accomplished animal behaviorist who had just published his seminal monograph on pronghorn. Dave was among a group of biologists beginning to integrate social behavior with ecology. He convinced me to spend time watching Roosevelt elk at Gold Bluffs Beach in northern California, helped me to understand elk behavior, and assisted with developing my critical thinking skills. My MS thesis concerned several aspects of elk behavior, but again, it was restricted to observations during rut. I realized that much of what we understood about animal behavior was not cast in an evolutionary perspective, and there were many unanswered questions concerning the links between ecology and behavior.

I did not become aware of sexual segregation until I was studying southern mule deer in the Cuyamaca Mountains of southern California for my doctoral dissertation at the University of Michigan (UM). My dissertation tested hypotheses about how resources and risk of predation helped shape the social organization of mule deer. My major professor, Dale R. McCullough, was studying the population ecology of white-tailed deer in southern Michigan when I arrived at the UM. After several years of rigorous coursework and thought-provoking discussions, I finally had my fieldwork well underway and was pursuing what I hoped would be cutting-edge science. There is no better preparation than having good mentors.

Dale visited my study site and subsequently sent me a draft of his new book on white-tailed deer. His book *The George Reserve Deer Herd* was to become the classic text on the population ecology of ungulates, and it reshaped our thinking about density-dependent processes and their role in population regulation and management. After 2 years at Michigan with Dale, including some work at the George Reserve, I felt prepared as a population ecologist. Tucked away in the latter part of Dale's book, however, was the chapter "Natural Selection and the Sexes." The concept of sexual segregation was completely new (at least to me), and I realized that the mule deer I was studying were also sexually separating outside the rut, but at a level far more pronounced than Dale had observed in white-tailed deer. In addition to my original hypotheses, I added potential factors that promoted sexual segregation to my list of objectives. Dale took a prominent position at the University of California before I finished my dissertation, and Gary E. Belovsky graciously oversaw the completion of my PhD, further refining my approach to scientific inquiry. In the end, I was firmly hooked on gaining a better understanding of sexual segregation.

I continue to ponder why the sexes of ungulates and other organisms live separately for a portion of the year. I have spent considerable time and effort (as have my students and colleagues) trying to discover the causes and consequences of sexual segregation. Many of my scientific endeavors lie at the interface of theory and application, and studying sexual segregation fits that pattern nicely. Sexual segregation has huge implications for the population dynamics, evolution of social behaviors, and conservation and management of ungulates. I have spent most of my career pursuing those ideas, working with colleagues and students at Unity College in Maine (6 years), the Institute of Arctic Biology (IAB) at the University of Alaska Fairbanks (UAF) (18 years), and Idaho State University (12 years). Although retired and now residing in northwest

Oregon, I have used my emeritus status at UAF (and an IAB appointment as a Senior Research Scientist) and interactions with my former colleagues and students to continue to offer new insights into sexual segregation. My goal in writing this book is to enhance our understanding of the causes and consequences of the separation of the sexes.

In this book, I integrate relevant information from a variety of organisms for comparative purposes and to help answer important questions. The book is intended for scientists and students of science, and it addresses questions of biology in a timely and logical manner—now is the time to sort among disjointed and sometimes muddled hypotheses concerning sexual segregation. Sexual segregation is widespread across organisms, and I believe that hunters, outdoor enthusiasts, naturalists, and anyone interested in why animals behave as they do will come away with an appreciation for a behavior that is not widely known or understood.

My writing style is scientific; I provide scientific names (genus and species) following the first use of common names for plants and animals so that there is no question about which species I am discussing. I also use citations to accord credit and provide evidence for statements of fact (and to offer the reader an opportunity to find additional literature on a particular topic). The articles cited in this book have undergone peer review (a rigorous and sometimes daunting process). Although imperfect, this process provides a level of credibility and confidence not afforded to nonscientific works. Unfortunately, scientific publications, and, hence, this book typically, are more difficult to read than the hunting and fishing magazines I enjoyed reading in my youth and that I still peruse. Those entertaining articles, which are typically narratives based on opinion, anecdotes, and sometimes experience, can contain inaccurate information because there are few checks and balances to ensure that information is presented accurately. The more challenging scientific style is the price paid for greater surety of knowledge. Nonetheless, I believe that anyone interested in natural history can decipher these writings to gain an additional appreciation of the natural world, and the behavior and ecology of the large mammals that roam across 6 continents. To that end, I have provided brief definitions of technical terms (at least with their first use) and US and metric equivalents for measurements (at least in the early chapters), and I offer "take home" summaries at the end of each chapter to help ease the process of reading a scientific work. Despite this, I concede that the layperson who jumps ahead to later chapters without first pursuing the introductory material may experience some rough sledding. Moreover, many hypotheses forwarded to explain sexual segregation in the latter chapters rely on details presented in the early ones.

My aim is straightforward but not simple. I propose to examine sexual segregation patterns, definitions, and hypotheses potentially explaining its occurrence. I will also explore the relevance of this phenomenon to theory and application, and I will clarify and organize the results into a coherent theme that will inform future research on this fascinating subject. In addition to scientists, I hope that this book is of interest to those who want to better understand the nature and behavior of these large, spectacular mammals.

Acknowledgments

I have spent much of my scientific career studying the ecology and behavior of large mammals. This book is a culmination of those efforts. I gratefully acknowledge the contributions of those scientists who preceded me and provided the intellectual foundations for this book. It is also my pleasure to acknowledge the colleagues and students who made this undertaking possible and exciting. To mention everyone who assisted me in this process, however, would be difficult, but I owe each a debt of deep gratitude. I was fortunate to have excellent colleagues at the institutions at which I taught and conducted research, and cadres of impressive undergraduates, graduate students, and postdoctoral associates to help, all of whom I would have difficulty thanking individually but nonetheless appreciate. I am especially grateful for the support I have received over the years from the Institute of Arctic Biology at the University of Alaska Fairbanks.

Charles R. Peterson contributed to my knowledge concerning sexual segregation in amphibians. Joel Berger suggested many improvements to the organization of this book; his research strongly influenced my understanding of sexual segregation. Several colleagues and former students deserve special thanks for their research on ungulates. John G. Kie helped develop ideas concerning resource partitioning between the sexes and applied his considerable skills in landscape ecology to our research. Perry S. Barboza was largely responsible for developing the gastrocentric model for sexual segregation and provided important insights into the physiology of ungulates. Vernon C. Bleich paved the way for understanding the role of predation risk in sexual segregation; he also provided numerous references on this topic. Becky M. Pierce helped refine my views of effects of predator size and hunting style on prey selection. Matthew C. Nicholson offered unique insights into migration of mule deer and provided a clear understanding of anything related to geographic information systems. Jeffrey T. Villepique integrated effects of predation risk with environmental variables. Kelley M. Stewart clarified how changing density of ungulates affected sexual segregation. Kevin L. Monteith developed new techniques for categorizing social groups of ungulates, and Jericho C. Whiting studied effects of water sources on sexual segregation in an arid ecosystem. Merav Ben-David provided a background in stable isotope analysis. Kris J. Hundertmark and Jennifer I. Schmidt kept me abreast of genetic advances related to large mammals, and Ryan A. Long increased my understanding of ungulate physiological ecology. Gail M. Blundell offered new insights into mammal social organization, and Patricia C. E. Reynolds help me better understand mammal ecology in the far north. David K. Person expanded my understanding of indices used to measure sexual segregation. Graduate students Douglas F. Spaeth, Susan A. Oehlers, Cody A. Schroeder, and Johanna C. Thalmann also conducted research that enhanced my knowledge of sexual segregation.

A number of individuals edited and provided helpful comments on portions of this book, including Carl D. Mitchell, Joel Berger, Janet L. Rachlow, Becky M. Pierce, Matthew C. Nicholson, Perry S. Barboza, Kelley M. Stewart, Kevin L. Monteith, Jeffrey T. Villepique, David K. Person, Vernon C. Bleich, Ryan A. Long, Jericho C. Whiting, and 2 anonymous reviewers. I am indebted to them all. Paul R. Krausman initially encouraged me to write this book, and Tiffany Gasbarrini, Ezra Rodriguez, and Kyle Kretzer oversaw the editorial process for Johns Hopkins University Press. Carrie Love did an outstanding job of copyediting, and Victor Van Ballenberghe graciously provided the cover photograph.

Finally, I am grateful for the support, patience, and love of my wife, Karolyn, and sons, Bryan and Jeffrey, and their tolerance of my extended absences during field seasons.

1 | Introduction and Overview

Historical Perspectives

What is sexual segregation? The scientific underpinnings for sexual segregation were first proposed by Charles Darwin (1859:93–94) in his discussion of sexual selection: *"Thus it is rendered possible for the two sexes to be modified through natural selection in relation to different habitats of life, as is sometimes the case; or for one sex to be modified in relation to the other sex, as commonly occurs."*

Darwin (1871) further clarified this line of reasoning by discussing the hypothesis of Boner (1861) for why the sexes of red deer (*Cervus elaphus*) lived apart for a portion of the year—males with antlers in velvet were less likely than females to inhabit densely wooded areas to avoid damaging those rapidly growing and sensitive structures. Darwin was partially incorrect (a rarity); instead of antlers, many ungulates (hooved mammals) that sexually segregate possess horns, which do not have velvet, and numerous species of open-land ungulates sexually segregate, as do many species inhabiting forests (Bowyer 2004). Darwin offered a proximate (immediate) observation to answer an ultimate question (why?), a long-standing difficulty in explanations for sexual segregation. Many ideas concerning sexual segregation have not stood the test of time (Miquelle et al. 1992, Main et al. 1996, Bleich et al. 1997, Bowyer 2004 for reviews); nonetheless, as with many evolutionary subjects, the foundation for this discipline clearly rests with Darwin (1859, 1871).

Those initial musings by Darwin led to widespread curiosity concerning why the sexes of some organisms do not live in close proximity to one another for a portion of the year. Over time, publications on sexual segregation have increased exponentially, especially for ungulates (Bowyer 2004). Nonetheless, sexual segregation remains among the more poorly understood, yet ardently debated, aspects of the behavior and ecology of plants and animals.

There are 2 general approaches for addressing hypotheses in ecology, behavior, and evolutionary biology. The first is to select the best organism (or assortment of related species) to examine a specific hypothesis so that there is a clear-cut and unequivocal outcome from the research. The second is to select an organism for which there is an extensive background of knowledge so that misunderstandings of biology and natural history are minimized. A study of sexual segregation in ungulates is well suited to incorporate both approaches, but a brief overview of ungulate biology is necessary to fully appreciate these complex mammals.

Ungulate Characteristics and Relevant Aspects of Their Behavior, Ecology, and Conservation

Before moving forward, I offer a more technical definition and additional background information for ungulates beyond just categorizing them as hooved mammals. This establishes context to review the ecology, behavior, and conservation of ungulates to set the stage for future discussions of sexual segregation.

Ungulates make up the mammalian order Perissodactyla, the odd-toed ungulates (31 species), including horses, zebras, asses, tapirs, and rhinoceroses. Ungulates also include terrestrial members of the order Cetartiodactyla (previously Artiodactyla), the eventoed ungulates (241 species). This taxonomic group, which is related distantly to whales and dolphins (Price et al. 2005), includes camels, peccaries, pigs, chevrotains, musk deer, muntjacs, giraffe, okapi, pronghorn, deer, hippos, and a diverse array of bovids (cattle, antelopes, sheep, bison, etc.) (Feldhamer et al. 2020).

Distribution, Habitats, and Body Size

Native species of ungulates are distributed widely across Africa, Asia, Europe, and North and South America (and have been introduced to Australia, New Zealand, and numerous islands); these large mammals inhabit climatic extremes, including arctic, temperate, and tropical environments. Ungulates occupy open habitats (savannahs and grasslands), ecotonal areas, closed tropical and temperate forests, and semiaquatic habitats; they occur from deserts to alpine mountain tops, and polar islands (Fritz and Loison 2006). Ungulates are among the most diverse of the mammalian radiations (Eisenberg 1981, Groves and Grub 2011).

Ungulates vary markedly in size, even for related species, such as African bovids, which range from the diminutive royal antelope (*Neotragus pygmaeus*), weighing 7–9 pounds (3–4 kg) and standing 10 inches (25 cm) at the shoulder, to the enormous giant eland (*Taurotragus derbianus*), which can weigh 1,000–2,000 pounds (455–907 kg) with a shoulder height of 69 inches (175 cm) (Dorst 1969). Ungulates are among the most sexually dimorphic of all mammals (Weckerly 1998b). Clinal variation exists in weights and *secondary sexual characteristics* (e.g., horns, pronghorns, antlers, and pedicles) related to body size and mating system (Loison et al. 1999, Pérez-Barbería et al. 2002).

Forest-dwelling species often are small, *monomorphic* (sexes of similar size), and lead a more solitary existence, whereas ungulates inhabiting open lands tend to be large, *sexually dimorphic* in size (with males larger than females), and gregarious. Smaller ungulates inhabiting closed forests and woodlands are more likely to be *monogamous* (having a single mate), with larger species living in more open habitats prone to be *polygynous* (one male mating with multiple females) (Estes 1974, Jarman 1983, Gosling 1986, Putman 1988).

Diets and Digestive Systems

Ungulates can consume a wide variety of plant parts and some animals. Pigs, warthogs (suids), and peccaries (tayassuids) consume some prey, but the bulk of their diet is composed of plants that are easy to digest. These "high-quality" items include tuberous roots, fallen fruit, herbaceous plants (including cacti), and the growing tips of grasses or sedges. Consequently, these omnivores have relatively simple digestive systems that are similar to those of humans (Stevens 1988). High-quality items tend to be patchily distributed and only available for brief periods of time. The more fibrous stems and leaves of woody plants (browse) and grasses are more difficult to digest but more abundant in space and time.

Other herbivorous ungulates have more complicated digestive systems that include chambers for digestion of plant fibers by microbial fermentation. Rhinos, horses, and tapirs save the difficult digestion in the cecum and proximal colon for last—readily digested sugars, starches, and proteins are removed before fermentation. This in-line fermentation system allows perissodactyls to efficiently remove starches from grasses and still digest some fiber at high intake (Demment and Van Soest 1985, Stevens 1988, Illius and Gordon 1991).

Ruminants (ungulates with a 4-chambered fermentation system that "chew their cud") account for most ungulate species. They are named for the rumen—a large vat in the front of the digestive system that allows storage and anaerobic fermentation of all the components of the diet. Microbial fermentation results in production of short-chain fatty acids that are used for energy. Ruminal fermentation provides an efficient extraction of fiber and a balanced supply of microbial protein, but intake of the long fibers of grasses may be limited by complex flow through the fermentation vat (Van Soest 1994). Large ruminants with broad muzzles and wide incisors are best suited to cropping grass swards and digesting the long fibers. Smaller ruminants with narrow muzzles and agile lips are better able to browse selectively and to digest the short, dense fiber fragments.

Ungulates can be categorized as browsers, intermediate feeders, or grazers (Hofmann 1985, 1988). African bovids that are browsers have narrow muzzles with large central and small lateral incisors designed for selective feeding on high-quality forage. Grazers, which feed on coarse forage, have broad muzzles with wide incisors of similar size adapted for nonselective feeding. A browser to grazer continuum of small to large body size occurs for those African bovids (Sinclair 2000), a pattern that holds for ungulates in general, with a few exceptions, such as moose (*Alces alces*), giraffe (*Giraffa camelopardalis*), and okapi (*Okapia johnstoni*), which are large browsers.

The capacity of the digestive system and its ability to hold and digest fiber increases with body size. Energy demands increase more slowly than gut capacity as animals get larger (Demment 1982). Large ungulates can afford to allocate more space and time to extracting energy from plant fiber. Con-

sequently, small ungulates may be unable to use abundant fibrous forages that are still acceptable to larger-bodied animals. Rare high-quality foods may be included in the diets of small and large ungulates when those items are abundant. Large grazing species can consume herbaceous plants and browse, but smaller species may be able to better satisfy their demands for energy and nutrients on these less abundant forages (Barboza et al. 2009, Craine 2021, Pardi and DeSantis 2021).

Hippos are unique in having a large stomach that allows for some pregastric fermentation of plants as well as possessing a comparatively low rate of energy metabolism—hippos are not especially efficient at digesting plant fiber, but their low energy demands do not require a high efficiency of nutrient extraction (Schwarm et al. 2006). Clearly, nutrition helps integrate behavioral and physiological responses of ungulates to the environments they inhabit (Parker et al. 2009). These relationships have relevance to sexual segregation and will be discussed further in Chapters 2 and 7.

Mating Systems

The diversity of ungulate mating systems is impressive (Bowyer et al. 2020a). These large mammals can exhibit territorial or nonterritorial behaviors for mating purposes. *Pair territories*, with both males and females often defending territory boundaries, occur among small forest-dwelling species, such as Kirk's dik-dik (*Madoqua kirkii*) and musk deer (*Moschus moschiferus*) (Brotherton and Manser 1997, Baskin and Danell 2003). Male ungulates also defend territories where polygyny is the primary system. *Polygynous resource territories* are usually defended by a dominant male and typically encompass important resources, such as food, water, or specific types of habitat. This system is the most common form of territoriality in polygynous species of African bovids (Estes 1974, Gosling 1986), but it also occurs in other species, such as pronghorn (*Antilocapra americana*) (Kitchen 1974). *Lek territories*, which often occur at traditional sites, are small and clustered with the most dominant male typically occupying the centrally located lek. These territories do not contain resources, except for the males themselves, which ostensibly are attractive to receptive females. Leks are known for topi (*Damaliscus lunatus*), kob (*Kobus kob*), fallow deer (*Dama dama*), and blackbuck (*Antilope cervicapra*) (Langbein and Thirgood 1989, Bro-Jørgensen 2002, Isvaran 2005, Ciuti and Apollonio 2008, Isvaran 2020). Bowyer et al. (2020a) provide a more complete review of leks, and hypotheses for their evolution.

Nonterritorial polygynous mating systems of ungulates include a *tending bond*, wherein a dominant males courts and defends a single female at a time; following copulation, the male leaves in search of additional mates (Bowyer et al. 2020a). Many mountain ungulates, such as bighorn sheep (*Ovis canadensis*) and mountain goats (*Oreamnos americanus*), exhibit tending bonds (Geist 1971, Festa-Bianchet and Côté 2008), as do mule deer (*Odocoileus hemionus*) and white-tailed deer (*O. virginianus*) (Kucera 1976, Hirth 1977, Bowyer 1986a, Airst and Lingle 2019). Similarly, bison (*Bison bison*) and caribou (*Rangifer tarandus*) engage in tending behaviors (Lent 1965, Berger and Cunningham 1994). The final nonterritorial system is a *harem*. A dominant male, the harem master, attempts to herd a group of females and defend that group from potential rivals. The male then endeavors to mate with females in the group as they come into estrus. Examples of ungulates that engage in harem mating include the muskox (*Ovibos moschatus*), red deer and North American elk (*Cervus elaphus*), wild reindeer (*Rangifer tarandus*), and wild horses (*Equus caballus*) (Clutton-Brock et al. 1982, Berger 1986, Bowyer and Kitchen 1987, Gunn 1992, Ihl and Bowyer 2011, Weladji et al. 2017).

Remarkably, species such as fallow deer, red deer, blackbuck, pronghorn, chamois (*Rupicapra rupicapra*), and wild horses may shift between types of mating systems (Rubenstein 1986, Byers and Kitchen 1988, Langbein and Thirgood 1989, Carranza et al.

1990, Duncan 1991, Isvaran 2005, Corlatti et al. 2013). Moreover, other species, such as moose and wild reindeer, can modify their mating system within a single rut (Bowyer et al. 2011, Weladji et al. 2017). These shifts in mating systems result from individual males altering their behaviors to take advantage of social and environmental changes to maximize their reproductive success (Bowyer et al. 2020a). Such variation in behavior results from *natural selection* (the differential reproduction of individuals) favoring the most reproductively successful individuals (Emlen and Oring 1977). Benefits that accrue from such behaviors above the level of the individual (i.e., to the population, species, or the environment) are neither necessary nor likely to explain such changes (Williams 1966, Barash 1982). I will discuss in Chapters 2 and 7 how sexual segregation can play a role in the mating behavior of ungulates, as well as the degree of sexual dimorphism in body size and the weaponry used in male-male combat.

Life-History Characteristics, Population Dynamics, and Movements

The life-history characteristics of ungulates related to sexual segregation are associated with ungulate population ecology. The life-history traits of large mammals differ markedly from those of small mammals (Caughley and Krebs 1983, Millar and Zammuto 1983), which means much of what we know concerning small mammals cannot be attributed to understand the ecology of large ones. Indeed, such differences form the foundation for assessing and evaluating the population dynamics of large mammals (McCullough 1979, Bowyer et al. 2014; Chapter 8). Large mammals possess numerous attributes consistent with a *slow-paced life history*, whereas small mammals (and many other small organisms) exhibit *fast-paced traits*. Thus, large mammals have comparatively slow development, have a delayed age at first reproduction, are *iteroparous* (give birth multiple times), possess small litters with large-bodied precocial young, are long lived, afford high maternal (but not paternal) investment to young, and exhibit a low *intrinsic rate of increase* (an innate ability to produce offspring, termed r) (Stearns 1977, Stubbs 1977, Gaillard et al. 2000). This suite of attributes promotes and responds to strong *density dependence* in large mammals (reproduction and survival are negatively associated with population density relative to the *ecological carrying capacity* of the environment, termed K) (Bowyer et al. 2014). Large body size buffers them against climatic extremes, and the slow-paced life history results in strong competitive abilities (Bowyer et al. 2020b). Ungulates with long lives may forgo or restrict allocation to reproduction to increase probability of their survival, or tradeoff current against future reproduction (Morano et al. 2013, Monteith et al. 2014b). Compelling evidence documents the widespread occurrence of density dependence among ungulates, especially for sexually dimorphic artiodactyls (McCullough 1979, Kie and White 1985, Fowler 1987, Boyce 1989, Stewart et al. 2005, Owen-Smith 2006, Bonenfant et al. 2009, Monteith et al. 2014b). In addition, ungulate populations may show irruptive behavior in the absence of sufficient harvest or predation, wherein there is a rapid increase in populations size, followed by an overshoot of K and a subsequent crash, resulting in negative consequences for their habitat (Leopold 1943, Klein 1968, McCullough 1997). I will return to how sexual segregation affects density dependence in ungulates in Chapter 8.

The *home range* (the areas to which an animal typically confines its daily activities) (Burt 1943) is positively related to body size in ungulates but varies with season, feeding type, sex, and distribution of resources (Mysterud et al. 2001). Some ungulates lead a relatively sedentary existence, whereas others make extensive and spectacular long-distance migrations, including those of 839 miles (1,350 km) for caribou, 480 miles (772 km) for mule deer, and 404 miles (650 km) for blue wildebeest (*Connochaetes taurinus*) (Joly et al. 2019). Ungulates exhibit variable patterns in whether they migrate or stop along their migratory paths (*stopovers*), the time required for mi-

grations, the distance traveled, and whether they are affected by anthropogenic developments, which holds import for their conservation and management (Fryxell and Sinclair 1988, Nicholson et al. 1997, Berger 2004, Monteith et al. 2011, Lendrum et al. 2012, 2013, 2014, Avgar et al. 2014, Sawyer et al. 2019). Ungulates have evolved an array of morphological, ecological, and behavioral patterns that are related to their life-history characteristics (Boyce 1988).

Predator Pits, Apparent Competition, and Territorial Regulation of Populations

Whether predation regulates ungulate populations remains a point of contention—the answer is of considerable importance to theoretical ecology and the management of these large mammals (Van Ballenberghe and Ballard 1994, Bowyer et al. 2005). Do predators regulate ungulate populations or simply track the dynamics of their primary prey? I briefly address this now, because questions concerning the role of sexual segregation in population dynamics of ungulates will come into play later (Chapter 8). For regulation to occur, there must be a density-dependent feedback (for instance, in prey as a result of predation or changes in population size as an outcome from harvest) (Holling 1959). Limitation requires only the death of individuals. Hence, when limiting factors operate in a density-dependent fashion, they are regulating and have the potential to maintain populations at densities lower than what the habitat would allow (Bowyer et al. 2014).

There are several theoretical mechanisms for predation regulating populations of ungulates at low density. One means of population regulation involves a *predator pit*. In a Ricker-like, stock-recruitment model (Ricker 1954, McCullough 1979) of population size (x axis; N_t) plotted against its size one reproductive effort later (y axis; N_{t+1}), there can exist a point of strong equilibrium occurring at low density (Fig. 1). This point results from predation causing the recruitment curve to dip below a 45° line of

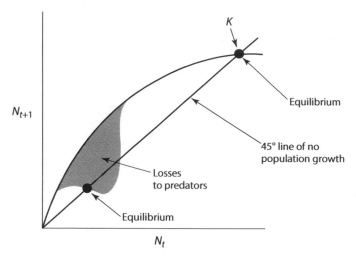

Fig. 1. A Ricker-like stock-recruitment graph illustrating the concept of a predator pit where predators can control ungulates at low density. Note the strong points of equilibria, and similar shape of the recruitment curve, at K (ecological carrying capacity) and at low density near the predator pit. This low point of equilibrium makes it difficult for prey to "escape" from the pit. (Courtesy of K. M. Stewart)

no growth (resulting in a pit). This low density of ungulates caused by intense predation represents a point of strong equilibrium (similar to population dynamics near K), making it difficult for populations to "escape" from the pit. Research on effects of predation by gray wolves (*Canis lupus*) and grizzly bears (*Ursus arctos*) on moose populations in Alaska and the Yukon offers strong support for this model (Gasaway et al. 1992). Clark et al. (2021) clarified the role of stochastic predation in exposing ungulate prey to predator-pit dynamics.

The types and effectiveness of predators and number of alternative prey may play a role in developing low-density equilibria; caution should be used in generalizing that population regulation of ungulates by their predators is commonplace. For example, predation by coyotes (*Canis latrans*) and mountain lions (*Puma concolor*) on mule deer slowed the rate of population growth but failed to control an increasing deer population (Pierce et al. 2012). Moreover, moose and wolf populations on Isle Royale have shown fluctuations rather than indications of low-density regulation of moose (Mech and Peterson 2003). Dale et al. (1994) concluded that the presence of multiple prey reduced the potential for wolves to regulate caribou numbers. Predator-prey systems may not achieve equilibrium because of environmental stochasticity, disease outbreaks, and a host of other factors. Clearly, there is no simple answer to the question of whether predators regulate their ungulate prey.

Another mechanism for predators to regulate prey involves *apparent competition*. This occurs when 2 *sympatric* (occurring in the same area) ungulates share a common predator. If one ungulate is abundant and the other rare, the greater number of predators supported by the more common species can adversely affect the population of the rarer ungulate, and give the appearance that the ungulates are competing (DeCesare et al. 2010, Wielgus 2017). For example, an endangered subspecies of bighorn sheep (*O. c. sierrae*) is preyed upon by mountain lions. The number of mountain lions, however, is reliant upon more numerous mule deer. More mountain lions occur than could be supported by numbers of bighorn sheep alone, and population dynamics of lions are mostly independent from those of bighorn sheep. Thus, the bighorn sheep population experienced higher rates of predation by mountain lions than if mule deer were at lower densities or absent (Johnson et al. 2013).

There is also a notion that territoriality among large carnivores regulates their population size independent of the number of ungulate prey, which has implications for predator-prey dynamics. An oft held opinion is that territorial aggression would preclude conspecifics from defended areas and thereby regulate numbers of predators, or that populations of territory holders would remain constant in size from antagonistic interactions resulting in mortality. I argue that the nature of territorial behavior and boundaries represents an adaptation by individuals to maximize their reproductive success; consequently, territorial (*contest*) competition alone does not offer a mechanism for population regulation of large carnivores. In theory, for territoriality to have evolved, territory holders must have been more successful than nonterritorial individuals that engaged in *scramble* (nonterritorial) competition (McCullough 1979, Pierce et al. 2000b; Fig. 2). Thus, simply precluding conspecifics from food via territorial behavior is not fundamentally different from removing food by scramble competition (Pierce et al. 2000b). Although more individuals might initially occupy an area under scramble than contest competition, more successful reproduction by territory holders would increase numbers beyond those of less suc-

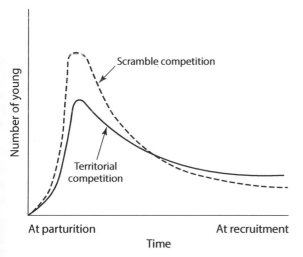

Fig. 2. Hypothesized relationship between outcomes of territorial (contest) and nonterritorial (scramble) competition on recruitment of young (modified from McCullough 1979). Although more individuals might initially breed and thereby produce more young under scramble than territorial competition, more young would be recruited if territoriality functioned to maximize reproductive success beyond that of individuals engaging in scramble competition. (From Pierce et al. 2000b)

cessful, nonterritorial individuals (McCullough 1979, Pierce et al. 2000b). Indeed, the factor ultimately regulating a population of territorial mammals is limitation of the resource that led to territorial behavior (Powell et al. 1997). Populations of large carnivores that exhibit territoriality are thought to be regulated primarily by the number and vulnerability of their ungulate prey (Schaller 1972, Bertram 1973, Macdonald 1983, Fuller 1989, Pierce et al. 2000a, b). This outcome has huge implications for the management of predator populations.

The size of wolf pack territories is inversely related to habitat quality (i.e., prey density)—more prey resulted in additional wolves with more but smaller territories. Either a constant number of wolves or their territories (Kittle et al. 2015, Sells et al. 2021) would be predicted if territoriality regulated populations. In addition, fluctuating numbers of large carnivores and their prey (Mech and Peterson 2003) produce a pattern inconsistent with territorial predators maintaining relatively constant populations resulting from self-regulation. The notion of self-regulation by predators lacks strong support in the scientific literature.

Effects on Plants, Animals, and Nutrient Cycling

Ungulates play a major role in affecting dynamics of ecosystems and can act as *keystone species*; that is, their presence and function has a disproportionate effect on other organisms in the ecosystem (Molvar et al. 1993, Simberloff 1998, Stewart et al. 2009). Large herbivores influence forage quality and availability through changes in plant production, composition of plant species, and rates and pathways of nutrient cycling (Bryant et al. 1983, Molvar et al. 1993, Pastor et al. 1993, Frank et al. 1994, Stewart et al. 2009, Guernsey et al. 2015). Indeed, *herbivore optimization* can occur when a peak in plant production ensues from moderate levels of herbivory compared with ungrazed or heavily grazed areas (McNaughton 1979a, Dyer et al. 1993, Stewart et al.

2006). The vast *grazing lawns* of the Serengeti offer strong evidence of such effects (McNaughton 1979b), although herbivore optimization is not limited to grasslands and can occur in other plant groups, including shrubs and forbs (Stewart et al. 2006). Low levels of herbivory can result in increased species richness and diversity of plants (Stewart et al. 2009), whereas high densities of ungulates can cause shifts in the species composition of entire forests (Risenhoover and Maass 1987). Further, foraging by ungulates can indirectly affect populations of insects, reptiles, birds, and small mammals (de Calesta 1994; Suominen et al. 1999a, b; Berger et al. 2001a; Flowerdew and Ellwood 2001; Keesing and Crawford 2001; McCauley et al. 2006; Berger et al. 2020). Clearly, large herbivores should be viewed as much more than consumers of plants and conveyors of urine and feces (Bowyer et al. 1997). The population dynamics of ungulates holds huge implications for ecosystems (Gordon and Prins 2019).

Trophic Cascades

Predators play a critical role in the behavior and ecology of ungulates (Terborgh et al. 2010), even when they do not regulate them. These large carnivores have captured our attention and stirred our imagination, sometimes provoking fear because of perceived or real threats to humans (Kruuk 2002). Modern field studies of large carnivores initially provided insights into their natural history and quantified their behavioral and ecological relationships with ungulates (Mech 1970, Kruuk 1972, Schaller 1972). Today, a multitude of scientific books deal with predators (Fox 1984, Carbyn et al. 1993, Caro 1994, Maehr 1997, Powell et al. 1997, Logan and Sweanor 2001, Creel and Creel 2002, Jenks 2018, Smith et al. 2020, to mention only a few), with some monographs casting these large mammals in a favorable light. Nonetheless, large carnivores tend to be either adored or vilified (Mech 2012, Dressel et al. 2015, Smith and Peterson 2021)—a result of their beauty and perceived ben-

efits to ecosystem health or their negative effects from killing livestock and native ungulates, as well as fear of the predators themselves.

Early researchers argued that large terrestrial herbivores seldom became food limited because predators held them at low numbers (well below *K*). This hypothesis, called HSS, is identified by the initials of the authors who proposed the concept (Hairston, Smith, and Slobodkin 1960). Simply put, *the world is green* because large herbivores are not limiting growth of vegetation. This process also is known as *top-down regulation*: predation by large carnivores, which are a trophic level above ungulates, holds ungulate populations below levels where they would have deleterious effects on the ecosystem. Conversely, when ungulate populations are larger and near *K*, food is limited, and regulation is termed *bottom-up*: ungulate populations are constrained by plants (forage), which are a trophic level below them (Bowyer et al. 2005, Pierce et al. 2012).

Categories of top-down and bottom-up regulation should not be viewed as simple dichotomies—both processes can operate in the same population over time, as Pierce et al. (2012) demonstrated for populations of mountain lions and mule deer. Moreover, simple trajectories of population increase and decline cannot be used to infer top-down or bottom-up regulation. A declining population of mule deer resulting from drought, and ostensibly overgrazing of their range, was attributed to bottom-up effects, whereas a slow increase in this population following drought was influenced by predation (Pierce et al. 2012). Nonetheless, top-down effects lead to the concept of *trophic cascades*, whereby the addition or removal of predators can result in profound changes to ecosystems (Terborgh et al. 2010). Sufficiently high densities of herbivores can drive successional pathways of plants and result in *ecological meltdowns* (Terborgh et al. 2001).

Trophic cascades are generally thought to be of 2 types: a numerical effect whereby predators hold prey at sufficiently low numbers to bring about changes in the ecosystem and behavioral cascades, where fear of predation alters prey behavior and thereby affects ecosystem processes (Berger et al. 2001b, Berger 2010, Ripple et al. 2016, Suraci et al. 2016). Debate continues over mechanisms causing trophic cascades (Kauffman et al. 2010, Mao et al. 2010, Ford et al. 2015, Painter et al. 2018, and many others), but there is no reason that numerical and behavioral processes cannot operate simultaneously. Persuasive evidence exists for behaviorally mediated trophic cascades in bushbuck (*Tragelaphus sylvaticus*) preyed upon by a suite of African carnivores (Atkins et al. 2019). Caro (2005) provides an excellent overview of changes in foraging behavior under risk of predation. This *landscape of fear* (Brown and Kotler 2004, Gaynor et al. 2019) plays a critical role in the behavior of mammals and has important implications for understanding sexual segregation. Nevertheless, care should be taken in assigning all observed behavioral responses to predation risk—not every environmental or behavioral change results from this process (Mech 2012, Villepique et al. 2015). Clearly, such environmental outcomes are intertwined in the mechanisms of population regulation for large mammals and hold important implications for sexual segregation.

Cultural and Historical Relationships with Ungulates

Why should we care about the ecology, behavior, and conservation of ungulates? I believe the answer entails more than the elegance of these large mammals, although beauty can be a compelling motivation. Human culture has been intertwined with ungulates and hunting throughout our evolutionary history, as evidenced by a 300,000-year-old throwing stick used by human ancestors in Germany, indications of hunting activities by early humans in Africa, and ~30,000-year-old cave paintings in France (Sadier et al. 2012, Bunn and Gurtov 2014, Conard et al. 2020). Paleolithic art, including petroglyphs of hunting scenes, and traditional use of ungulates by Indigenous Peoples of North America speak to

how completely Indigenous cultures and ungulates are interwoven (e.g., Murie 1935, Allen 1954, McHugh 1972, Guthrie 2000, McCabe 2002). We are mammals; I believe that this connection promotes inquisitiveness, stimulates our imagination, and fosters empathy for the other large mammals with which we share the Earth. Moreover, our close and long-term association with wild and domestic ungulates for food and sport has fostered our appreciation of these large mammals, thereby enhancing their intrinsic value (Bowyer et al. 2019). Indeed, the ancient Greeks engaged in sport hunting over 2 millennia ago (Hull 1964), a practice that continues throughout the modern world and is epitomized by highly regulated hunting in North America, Europe, and parts of Africa (Bowyer et al. 2019).

Historic declines of wildlife in Europe and North America were pronounced (Linnell and Zachos 2011, Bowyer et al. 2019). In North America, declines during the 1800s resulted from excessive harvests, some of which were tied to economic incentives and commercial markets (Geist 1988). The following 2 centuries involved considerable effort to restore and conserve wildlife populations in North America (Leopold 1933, Allen 1954, Trefethen 1975, Bowyer et al. 2019). Declines in numbers of whitetailed deer, North American elk, pronghorn, and American bison were especially prominent; bison numbers waned principally as a result of war waged against Indigenous Peoples by the United States, which included attempting to eliminate their food supply (Allen 1954). By the late 1800s, declining wildlife populations prompted Theodore Roosevelt and a select group of hunters and naturalists to form the Boone and Crockett Club to initiate major conservation initiatives to restore populations of ungulates in North American (Organ et al. 2010, Krausman and Bleich 2013). Hunting regulations with restricted seasons and quotas were implemented in many localities by the early 1900s, although regulation of hunting in some localities had been enacted much earlier (Cummings 2019). Moreover, reintroduction programs were initiated to restore depleted or extirpated populations, including those of bighorn sheep (Bleich et al. 2018, Robinson et al. 2019). Large carnivores likewise have been restored to parts of their previous distribution, but they are still persecuted in some areas (Ripple et al. 2014). Nonetheless, conservation measures were effective for numerous large mammals; today, in North America and Europe, wild populations of mammals are prospering, and legal hunting is allowed for a number of species (Organ et al. 2010, Bowyer et al. 2019, Mahoney and Geist 2019). Moreover, hunters in North America have contributed billions of dollars toward the conservation and management of wildlife (Southwick and Allen 2010, Heffelfinger and Mahoney 2019). Simply put, there would be few free-ranging populations of ungulates in North American to enjoy or study without those long-term efforts.

Challenges Today and Future Unknowns

There are other reasons to care about mammals. Overall, mammal populations, including ungulates, are imperiled worldwide, as indicated by the International Union for Conservation of Nature's *IUCN Red List of Threatened Species* (Bowyer et al. 2019). Rates of recent extinctions far exceed historical levels, requiring increased efforts to conserve threatened mammals (Ceballos and Ehrlich 2002, Isaac et al. 2007, Pimm et al. 2014, Ceballos et al. 2017). Slightly over 85% of the 27 orders of the world's mammals exhibit some level of threat to their existence (Bowyer et al. 2019; Fig. 3), a profound challenge for conservation biologists. Orders with large-bodied species, including ungulates (Cetartiodactyla and Perissodactyla), have especially high threat levels (Schipper et al. 2008), which is a mea-

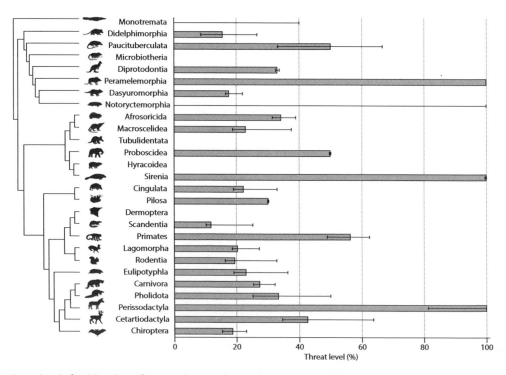

Fig. 3.   Threat levels for 27 orders of mammals. Error bars indicate lower and upper bounds for threat levels. Threat level = [(VU + EN + CR + EW)/(Total − DD)] × 100. The lower bound for the threat level = [(VU + EN + CR + EW)/Total] × 100; the upper bound for the threat level = [(VU + EN + CR + EW + DD)/Total] × 100. Abbreviations represent number of species in IUCN Red List category. VU = Vulnerable; EN = Endangered; CR = Critically Endangered; EW = Extinct in the Wild; DD = Data Deficient. J. Kenagy and J. Bradley provided the phylogeny for inclusion in this figure. (From Bowyer et al. 2019)

sure of their peril (Fig. 3). In addition to taxonomic order and body size, other intrinsic characteristics—such as a small geographic range, inhabiting islands rather than the mainland, and having a slow-paced life history—are associated with high risks of extinction (Davidson et al. 2017). These characteristics of large mammals also hold the potential to increase susceptibly to harvest by humans (Cardillo et al. 2005). Those same life-history traits that make mammals vulnerable to extinction also increase their risk to adverse effects from a changing climate (Davidson et al. 2017). Ungulates are not exempt from such risks (Bowyer et al. 1998b, Forchhammer et al. 2001, Lenart et al. 2002, Post and Forchhammer 2008, Cedir et al. 2015, Montcith et al. 2015, Berger et al. 2018, Mallory and Boyce 2018). Indeed, a rapidly changing climate accelerates risks of extinction for mammals, and it bodes poorly for the continued existence of many species (Urban 2015).

Deforestation threatens terrestrial mammals, including ungulates, throughout the tropics, especially in the Americas, Asia, and Africa (Schipper et al. 2008). Negative effects on land mammals of illegal killing are most pronounced in Asia, but are also apparent in parts of Africa and South America. Over-harvesting may result from poaching for food, the bushmeat trade, and use of animals for medicinal and other purposes, as well as from accidental capture in snares set for other species (Schipper et al. 2008, Harrison et al. 2016, Bowyer et al. 2019). Indeed, unregulated or illegal killing of wild mammals by humans continues to be a threat to some mammals, especially in tropical regions with diverse mammalian faunas (Van Vliet et al. 2015). Yet, in North America, Europe, and parts of Africa, hunting has formed the basis for successful programs to ensure conservation of critical habitats and to restore wild populations (Organ et al. 2010, Bowyer et al. 2019, Mahoney and Geist 2019). Resilience of populations to hunting, driven by density-dependence processes, is a potent stabilizing force that ensures sustainable harvests so long as hunting is properly managed. Regrettably, un-

regulated or illegal killing continues for cultural or economic reasons worldwide (Bowyer et al. 2019). Overcoming these barriers to conservation is difficult—but necessary to prevent the continued loss of many mammals, including ungulates. Unmistakably, the ultimate fate of ungulates and the ecosystems they inhabit will be determined by the size, growth, and resource demands of the human population (Vitousek et al. 1997, McKee et al. 2013. Bowyer et al. 2019, Berger et al. 2020).

Traditionally, protected areas, such as parks, reserves, or other locales intended to limit human activities, have been the foundation for conservation efforts to sustain mammals and other wildlife. Recently, however, conservationists have realized that protected areas around the world often are too small, too few, too poorly delimited, too isolated, or too unreliably supported to meet many long-term needs for safeguarding mammals (Bowyer et al. 2019 for review). The size of protected areas and demography of humans are critical factors influencing species richness and risk of extinction (Brashares et al. 2001). For example, the size of protected areas in East Africa is a chief determinant of species richness for ungulates (Bowyer et al. 2019). Opportunities for creating new conservation areas, however, are restricted by a changing climate and increasing demand on land use by a growing human population (Payne and Bro-Jørgensen 2020). Although such areas form an important foundation for the conservation of ungulates, in the long run, they are not sufficient. Conservation efforts need to be extended and enhanced in human-occupied landscapes, where multiple land uses occur, but with a renewed emphasis on better integrating human concerns in prioritizing conservation to maintain the viability of wildlife populations over time. Nonetheless, factors that threaten the existence of numerous species cannot be eliminated entirely, and those mammals will need to be managed—not simply protected—to ensure their persistence; such species are *conservation reliant* (Goble et al. 2012, Scott et al. 2020). Long-term conservation of many, if not most, wild mammals, including ungulates, will require more intensive efforts.

Sexual Segregation in Plants and Animals

I now turn to the topic of the book—sexual segregation in ungulates. Separation of the sexes occurs across a wide array of plants and animals (Table 1), and many species will be added to this list as more studies are undertaken. Excellent reviews of vertebrates by Ruckstuhl and Neuhaus (2005) and Wearmouth and Sims (2008) contain many citations that I did not include—for the most part, I have not duplicated citations for sexual segregation in those reviews but have concentrated on adding more recent papers. Rarely, I have included citations that were overlooked. I also have added review papers for other taxonomic groups, which should serve as starting points for those interested in broader treatments of sexual segregation. For ungulates, however, I have compiled a more complete list to assist with forthcoming discussion of reasons underpinning sexual segregation. Nonetheless, I have restricted my search to peer-reviewed articles; consequently, government reports, master's theses, and doctoral dissertations are not included.

I undoubtedly missed papers on sexual segregation—some likely by constraining my search with particular keywords (e.g., sexual segregation, differences between the sexes), and others as a result of my lack of expertise with some taxa in Table 1. Moreover, I did not assess the types of, or evidence for, sexual segregation in those publications—papers were included based on determinations made by their authors. I have held off citing publications that deal with assessing and understanding differences between the sexes of ungulates until a broader consideration of hypotheses in Chapter 5–7.

My arrangement of organisms in Table 1 unquestionably will puzzle some taxonomists; I did not follow a classic *phylogenetic scheme* (the evolutionary history and relationships among taxa). I was more interested in grouping taxa based on similar

life-history characteristics that represented ecological equivalents of ungulates. For instance, *macropods* (kangaroo, wallaby, etc.) have many attributes that are similar to the lifestyles of ungulates (McCullough and McCullough 2000), as do elephants (large body size, slow-paced life histories, herbivory, etc.). *Cetaceans* (whale and dolphins) are relatives of artiodactyls, but their ecology has more in common with seals, especially the *odontocetes* (toothed whales), because of their aquatic and carnivorous lifestyles. Bats, because of flight, have obvious ecological connections to most birds. Nevertheless, I suspected that *ratites* (flightless birds, such as the ostrich, emu, rhea, and cassowary) might also lead a somewhat ungulate-like existence, but I located no evidence in the literature that they sexually segregate.

Despite my caveats on the number of species in Table 1, there are some surprising and unexpected outcomes. I did not previously realize that at least one parasite exhibited sexual segregation within the body of its host (Tuomainen et al. 2015). There was also little relation between the number of described species within a taxon and the number of examples of sexual segregation in the published literature. I recognize that many undescribed species likely exist, especially for some taxa, which might further affect this relationship. Remarkably, I located only 11 publications on sexual segregation in invertebrates (Table 1), yet there are ~950,000 species of insects alone (Mora et al. 2011). I am unsure if this results from an absence of sexual segregation in those numerous organisms, or if entomologists simply are not interested in that aspect of insect behavior and ecology because of other research priorities. Reptiles (~8,200 species) are well represented in Table 1 if references in the review paper by Shine and Wall (2005) are considered. Nonetheless, I located relatively few publications on sexual segregation in amphibians (frogs, newts, salamanders, etc.) (~6,200 species).

Among other vertebrates, fishes are well represented in Table 1; considering there are ~30,000 described species, this is not surprising. Birds (~10,000

Table 1. Selected publications on sexual segregation in plants and animals

Plants
    Dioecious plants (reviews)   Freeman et al. (1976), Bierzychudek and Eckhart (1988), Harder et al. (2000), Barrett and Hough (2013), Van Drunen and Dorken (2014), Torices et al. (2019)

    Sorrel (*Rumex acetosa, R. acetosella*)   Korpelainen (1991)
    Bluegrass (*Poa ligularis*)   Bertiller et al. (2002)
    Saltgrass (*Distichlis spicata*)   Mercer and Epplex (2010)
    Pines (*Pinus* spp.)   Shmida et al. (2000)
    English yew (*Taxus baccata*)   Garbarino et al. (2015)
    Desert moss (*Syntrichia caninervis*)   Stark et al. (2005)

Insects
    Cipero's ground hopper (*Tetrix ceperoi*)   Hochkirch et al. (2007)
    Whirligig beetle (Coleoptera: Gyrinidae)   Romey and Wallace (2007)
    Milkweed beetles (*Tetraopes tetrophthalmus, T. femoratus*)   Lawrence (1982)
    Anthocorid (*Orius sauteri*)   Nakashima and Hirose (2003)
    Chinese white wax scale (*Ericerus pela*)   Zhao et al. (2013)
    Fender's blue butterfly (*Plebejus icarioides fenderi*)   Thomas and Schultz (2016)

Spiders
    Harvestmen (*Prionostemma* spp.)   Harvey et al. (2017)

Other Invertebrates
    Squid (*Loligo gahi*)   Arkhipkin and Middleton (2002)
    Common octopus (*Octopus vulgaris*)   Alonso-Fernádez et al. (2017)
    Leach (*Aegla* spp.)   Barría et al. (2011)
    Helminth (*Echinorhynchus borealis*)   Tuomainen et al. (2015)

Fishes
    Marine fishes (review)   Wearmouth and Sims (2008)
    Sharks (reviews)   Sims (2005), Mucientes et al. (2009), Braccini and Taylor (2016), Drymon et al. (2020)
    Spiny dogfish (*Squalus acanthias*)   Dell'Apa et al. (2014), Haugen et al. (2017)
    Whale shark (*Rhincodon typus*)   Ketchum et al. (2013)
    Great white shark (*Carcharodon carcharias*)   Robbins (2007), Domeier and Nasby-Lucas (2012), Kock et al. (2013), Milankovic et al. (2021)
    Hammerhead shark (*Sphyrna lewini*)   Klimley (1987)
    Broadnose sevengill shark (*Notorynchus cepedianus*)   Barnett et al. (2011)
    Spotted catshark (*Scyliorhinus canicula*)   Wearmouth et al. (2012)
    Redspotted catshark (*Schroederichthys chilensis*)   Vásquez-Castillo et al. (2021)
    Skates (Rajiformes)   Bizzarro et al. (2014)
    Skates (*Raja microocellata, R. brachyura, R. montagui, R. clavata*)   Simpson et al. (2021)
    Chimaeras (*Chimaera* spp., *Hydrolagus* spp.)   Holt et al. (2013)
    Trinidadian guppy (*Poecilia reticulata*)   Croft et al. (2004, 2006)
    European minnow (*Phoxinus phoxinus*)   Griffiths et al. (2014)
    Three spine stickleback (*Gasterosteus aculeatus*)   Ruckstuhl (2007)
    Atlantic and sailfin molly (*Poecilia Mexicana, P. latipinna*)   Scharnweber et al. (2011)
    Zebra fish (*Danio rerio*)   Aivaz and Ruckstuhl (2011)
    European cyprinids (*Rutilus, Abramis, Alburnus*)   Žák et al. (2020)

Reptiles
    Crocodilians, turtles, lizards, snakes (review)   Shine and Wall (2005)
    Marine reptiles (review)   Wearmouth and Sims (2008)
    Graceful crag lizard (*Pseudocordylus capensis*)   Eifler et al. (2007)
    Brown anole lizard (*Anolis sagrei*)   Delaney and Warner (2016)
    Map turtle (*Graptemys geographica*)   Bulté et al. (2008)

Amphibians
    Tusked frog (*Adelotus brevis*)   Katsikaros and Shine (1997)
    Columbia spotted frog (*Rana luteiventris*)   Pilliod et al. (2002)

(continued)

Table 1. (*continued*)

Brilliant-thighed poison frog (*Allobates femoralis*)   Ringler et al. (2009)
Green and golden bell frog (*Litoria aurea*)   Valdex et al. (2016)
Western toad (*Bufo boreas*)   Bartelt et al. (2004)
Mole salamander (*Ambystoma talpoideum*)   Groff et al. (1981)
Paghman Mountain salamander (*Batrachuperus mustersi*)   Reilly (1983)
Golden-striped salamander (*Chioglossa lusitanica*)   Sequeira (2001)
Alpine newt (*mesotriton alpestris*)   Kopecký et al. (2012)
Italian crested newt (*Triturus carnifex*)   Romano et al. (2012)

Birds

Penguins, albatross, petrels, boobies, ducks, raptors, grouse,   Catry et al. (2005)
   shorebirds, shrikes, robins, warblers, kinglets, buntings (review)
Near-Arctic migrants (review)   Komar et al. (2005)
Marine birds (reviews)   Wearmouth and Sims (2008), Phillips et al. (2011)
Migratory land birds (review)   Bennett et al. (2019)
Migrant warblers (*Setophaga ruticilla, S. caerulescens, S. discolor*)   Spidal and Johnson (2016)
Wood warblers (*Dendroica magnolia, D. coronate, D. virens, D. fusca*)   Morse (1968)
Black-and-white warbler (*Mniotilta varia*)   Cooper et al. (2021)
American redstart (*Setophaga ruticilla*)   Marra and Holmes (2001)
Henslow's sparrow (*Ammospiza henslowii*)   Robins (1971)
Eurasian skylark (*Alauda arvensis*)   Powolny et al. (2016)
Chiffchaff (*Phylloscopus collybita*)   Catry et al. (2007)
Willow ptarmigan (*Lagopus lagopus*)   Elson et al. (2007)
Black grouse (*Tetrao tetrix*)   Ciach et al. (2010)
Pheasant (*Phasianus colchicus*)   Hill and Ridley (1987), Whiteside et al. (2017, 2018, 2019),
                  Pallante et al. (2021)
Greater sage-grouse (*Centrocercus urophasianus*)   Smith et al. (2018)
Capercaillie (*Tetrao urogallus*)   Bañuelos et al. (2008)
Lesser bustard (*Tetrax tetrax*)   Morales et al. (2008)
Great bustard (*Otis tarda*)   Palacín et al. (2009), Alonso et al. (2016), Bravo et al. (2016)
Lesser kestrel (*Falco naumanni*)   Catry et al. (2016)
American kestrel (*Falco sparverius*)   Smallwood (1987)
Andean condor (*Vultur gryphus*)   Perrig et al. (2021)
Magellanic woodpecker (Campephilus magellanicus)   Duron et al. (2017)
Imperial shag (*Phalacrocorax atriceps*)   Harris et al. (2013)
Yellow-legged gulls (*Larus michahellis*)   Calado et al. (2020)
Northern and southern royal albatrosses (*Diomedea sanfordi*,   Jiménez et al. (2017)
   D. epomophora)
Wandering albatross (*Diomedea exulans*)   Pereira et al. (2018)
Black-browed albatross (*Thalassarche melanophris*)   Paz et al. (2021)
Monteiro's storm petrel (*Hydrobates monteiroi*)   Paiva et al. (2018)
Giant petrels (*Macronectes halli, M. giganteus*)   Forero et al. (2005), González-Solís and Croxall (2005),
                  Granroth-Wilding and Phillips (2019)
Snow petrel (*Pagodroma nivea*)   Barbraud et al. (2019)
Sooty shearwater (*Puffinus griseus*)   Hedd et al. (2014)
Shearwaters (*Calonectris Diomedea, C. borealis, C. edwardsii*)   De Felipe et al. (2019)
Black skimmer (*Rynchops niger*)   Mariano-Jelicich et al. (2007)
Northern gannet (*Morus bassanus*)   Stauss et al. (2012), Cleasby et al. (2015), Clark et al. (2021)
Scopoli's shearwater (*Calonectris diomedea*)   Reyes-González et al. (2021)
Red-footed booby (*Sula sula*)   Weimerskirch et al. (2006)
Brown booby (*Sula leucogaster*)   Miller et al. (2018), Austin et al. (2021)
Masked booby (*Sula dactylatra*)   Lema (2020)
Black-tailed godwit (*Limosa limosa*)   Catry et al. (2012)
Rockhopper penguin (*Eudyptes chrysocome*)   Ludynia et al. (2013)
Gentoo penguin (*Pygoscelis papua*)   Xavier et al. (2017), Tian et al. (2021)
Magellanic penguin (*Spheniscus magellanicus*)   Barrionuevo et al. (2020)

MAMMALS

Bats

Bats (reviews)

Kunz and Lumsden (2003), Altringham and Senior (2005), Senior et al. (2005), Angell et al. (2013)

Noctule bats (*Nyctalus noctule, N. leisleri, N. lasiopterus*) Ibáñez et al. (2009)
Little brown bat (*Myotis lucifugus*) Burns and Broders (2015)
Daubenton's bat (*Myotis daubentonii*) Encarnãço (2012), Cistrone et al. (2015), Nardone et al. (2015), Linton and Macdonald (2019)

Bechstein's bat (*Myotis bechsteinii*) Katsis et al. (2021)
Natter's bat (*Myotis nattereri*) Katsis et al. (2021)
Brown long-eared bat (*Plecotus auritus*) Katsis et al. (2021)
Greater mouse-tailed bat (*Rhinopoma microphyllum*) Levin et al. (2013)
Western barbastelle bat (*Barbastella barbastellus*) Hillen et al. (2011)
Parti-colored bat (*Vespertilio murinus*) Safi (2008), van Toor et al. (2011)
Evening bat (*Nycticeius humeralis*) Istvanko et al. (2016), Kwon et al. (2019)
Hoary bat (*Lasiurus cinereus*) Cryan and Wolf (2003)
Pteropodid bat (*Otopteropus cartilagonodus*) Ruedas et al. (1994)
Sheba's short-tailed bat (*Carollia perspicillata*) Alvis and Pérez-Torres (2020)

Seals, Sea Lions, and Walrus

Seals (reviews)

Staniland (2005), Wearmouth and Sims (2008), Krüger et al. (2014)

Fur seals (*Arctocephalus gazella, A. tropicalis*) Kernaléguen et al. (2012)
Fur seals (*Arctocephalus gazella, A. tropicalis, A. australis*) de Albernaz et al. (2016)
Antarctic fur seal (*Arctocephalus gazella*) Staniland and Robinson (2008), Kernaléguen et al. (2016), Jones et al. (2020a, b)

South American fur seal (*Arctocephalus australis*) de Lima et al. (2019)
Australian fur seal (*Arctocephalus pusillus*) Kernaléguen et al. (2015), Salton et al. (2019)
Harbor seal (*Phoca vitulina*) Kovacs et al. (1990)
Weddell seal (*Leptonychotes weddellii*) Langley et al. (2018)
Gray seal (*Halichoerus grypus*) Beck et al. (2003, 2007), Tucker et al. (2007)
Hooded seal (*Cystophora cristata*) Bajzak et al. (2009)
Southern seal lion (*Otaria flavescens*) Drago et al. (2006), Baylis et al. (2016)
Galápagos sea lion (*Zalophus californianus wollebaeki*) Wolf et al. (2005)
Northern elephant seal (*Mirounga angustirostris*) Gomez and Cassini (2014)
Southern elephant seal (*Mirounga leonine*) Lewis et al. (2006), McIntyre et al. (2010)
Walrus (*Odobenus rosmarus*) Fay (1982)

Whales, Dolphins, and Porpoises

Toothed whales (review) Michaud (2005)
Whales and dolphins (review) Wearmouth and Sims (2008)
Minke whale (*Balaenoptera acutorostrata*) Laidre et al. (2009)
Killer whale (*Orcinus orca*) Baird et al. (2005), Beerman et al. (2015)
Sperm whale (*Physeter macrocephalus*) Pirotta et al. (2020)
Australian humpback dolphin (*Sousa sahulensis*) Hawkins et al. (2020)
Bottlenose dolphin (*Tursiops truncates*) Morteo et al. (2014)
Indo-Pacific bottlenose dolphin (*Tursiops aduncus*) Fury et al. (2013), Sprogis et al. (2016, 2018), Galezo et al. (2018)

Franciscana dolphin (*Pontoporia blainvillei*) Danilewicz et al. (2009)
Hector's dolphin (*Cephalorhynchus hectori*) Webster et al. (2009)
Beluga (*Delphinapterus leucas*) Loseto et al. (2006), Szpak et al. (2019)
Dusky dolphin (*Lagenorhynchus obscurus*) Shelton et al. (2010)

Terrestrial Carnivores

Mountain lion (*Puma concolor*) Keehner et al. (2015), Johansson et al. (2018)
Bobcat (*Lynx rufus*) McNitt et al. (2020)
European wildcat (*Felis silvestris silvestris*) Oliveira et al. (2018)
Jaguar (*Panthera onca*) Conde et al. (2010), Gese et al. (2018)
Snow leopard (*Panthera uncia*) Johansson et al. (2018)
Serval (*Leptailurus serval*) Ramesh et al. (2015)

(continued)

Table 1.   (*continued*)

Grizzly bear (*Ursus arctos*)	Wielgus and Bunnell (1995)
Brown bear (*Ursus arctos*)	Rode et al. (2006), Steyaert et al. (2013)
Black bear (*Ursus americanus*)	Gantchoff et al. (2019)
Cave bear (*Ursus spelaeus*)[extinct]	Germonpré (2004)
Giant panda (*Ailuropoda melanoleuca*)	Qi et al. (2011)
European polecat (*Mustela putorius*)	Lodé (1996), Marcelli et al. (2006)
Pine marten (*Martes martes*)	Zalewski (2007)
Wolverine (*Gulo gulo*)	van Dijk et al. (2008)
Sea otter (*Enhydra lutris*)	Elliott Smith et al. (2015)
Narrow-striped mongoose (*Mungotictis decemlineata*)	Schneider and Kappeler (2016)
Hares, Rabbits, and Pikas	
Snowshoe hare (*Lepus americanus*)	Litvaitis (1990), Ellsworth et al. (2016)
Brown hare (*Lepus europaeus*)	Husek et al. (2015)
Mountain hare (*Lepus timidus*)	Rehnus and Bollmann (2020)
Rodents	
Deer mouse (*Peromyscus maniculatus*)	Bowers and Smith (1979)
White-footed mouse (*Peromyscus leucopus*)	Morris (1984)
Meadow vole (*Microtus pennsylvanicus*)	Morris (1984), Morris and MacEachern (2010)
Eurasian beaver (*Castor fiber*)	Lodberg-Holm et al. (2021)
Hispid cotton rat (*Sigmodon hispidus*)	Cameron and Spencer (2008)
Primates	
Nonhuman primates (reviews)	Clutton-Brock (1977), Watts (2005)
Green monkey (*Cercopithecus sabaeus*)	Harrison (1983)
Spider monkey (*Ateles geoffroyi*)	Hartwell et al. (2014, 2018, 2021)
Bornean orangutan (*Pongo pygmaeus*)	Schuppli et al. (2021)
Human (*Homo sapiens*) (reviews)	Pellegrini et al. (2005), Alves (2020)
Elephants	
African elephant (*Loxodonta africana*)	Owen-Smith (1988), Poole (1994), Stokke (1999), Stokke and du Toit (2000, 2002, 2014), Shannon et al. (2006a, b; 2008), Smit et al. (2007), Kioko et al. (2020)
Asian elephant (*Elephas maximus*)	Sukumar and Gadgil (1988), Sukumar (1989)
Australian Marsupials	
Honey possum, mountain pygmy-possum, western gray kangaroo, white-fronted dunnart, agile antechinus, northern quoll (reviews)	MacFarlane and Coulson (2005), Garnick et al. (2014)
Western gray kangaroo (*Macropus fuliginosus*)	Coulson et al. (2006), MacFarlane (2006), MacFarlane and Coulson (2007, 2009), Garnick et al. (2018)
Swamp wallaby (*Wallabia bicolor*)	Di Stefano et al. (2009)
Ungulates	
Red deer (*Cervus elaphus*)	Boner (1861), Darling (1937), Charles et al. (1977), Staines (1977), Watson and Staines (1978), Clutton-Brock et al. (1982a, 1987), Staines et al. (1982), Clutton-Brock and Albon (1989), Conradt (1998a, b), Conradt et al. (1999, 2000, 2001), Conradt and Roper (2000), Bonenfant et al. (2004), Siuta (2006), Kamler et al. (2007), Zweifel-Schielly et al. (2011), Azorit et al. (2012), Bocci et al. (2012), Miranda et al. (2012), Alves et al. (2013, 2014), Debeffe et al. (2019)
North American elk (*Cervus elaphus*)	Peek and Lovas (1968), Flook (1970), Geist (1982), Weckerly (1998b, 2001, 2017), Maier and Post (2001), Weckerly et al. (2001, 2004), Walter (2014), Gregory et al. (2009), Long et al. (2009b), Vander Wal et al. (2013), Stewart et al. (2015), Bliss and Weckerly (2016), Peterson and Weckerly (2017), O'Brien et al. (2018)
Sika (*Cervus nippon*)	Koga and Ono (1994), Jiang et al. (2008)
Eld's deer (*Cervus eldi*)	McShea et al. (2001), Yan et al. (2013)

Père David's deer (*Elaphurus davidianus*)	Jiang et al. (2000)
Pampas deer (*Ozotoceros bezoarticus*)	Cosse and González (2013)
Huemul (*Hippocamelus bisulcus*)	Frid (1994)
White-tailed deer (*Odocoileus virginianus*)	McCullough (1979), Ozoga et al. (1982), Marchinton and Hirth (1984), Beier (1987), Verme (1988), McCullough et al. (1989), Beier and McCullough (1990), Weckerly and Nelson (1990), LaGory et al. (1991), Nixon et al. (1991), Brockmann and Pletscher (1993), Jenks et al. (1994), Leslie et al. (1996), Kie and Bowyer (1999), Lesage et al. (2002), DePerno et al. (2003), Stewart et al. (2003b, 2011), Fulbright and Ortega-S. (2006), Zimmerman et al. (2006), Monteith et al. (2007), Richardson and Weckerly (2007), Weckerly (2010), Dechen Quinn et al. (2013), Haus et al. (2020)
Mule deer and black-tailed deer (*Odocoileus hemionus*)	King and Smith (1980), Bowyer (1984), Ordway and Krausman (1984), Scarbrough and Krausman (1988), Weckerly (1993), Bowyer et al. (1996), Main and Coblentz (1996), Nicholson et al. (1997), Bowyer and Kie (2004), Farmer et al. (2006), Zimmerman et al. (2006), Shields et al. (2012), Gallina-Tessaro et al. (2019), Rodgers et al. (2021)
Fallow deer (*Dama dama*)	Putman et al. (1993), Thirgood (1996), Focardi et al. (2003), Ciuti et al. (2004), Apollonio et al. (2005), Villerette et al. (2006a), Ciuti and Apollonio (2008), Azorit et al. (2012)
Reindeer (*Rangifer tarandus*)	Helle (1979), van Wieren and de Bie (1979), Skogland (1989), Loe et al. (2006)
Caribou (*Rangifer tarandus*)	Cameron and Whitten (1979), Jakimchuk et al. (1987), Barboza et al. (2018)
Moose (*Alces alces*)	Edwards (1983), Miller and Litvaitis (1992), Miquelle et al. (1992), Bowyer et al. (2001b), Spaeth et al. (2001, 2004), Dussault et al. (2005), Olsson et al. (2010), Oehlers et al. (2011), Winkler and Kaiser (2011), Bjøneraass et al. (2012), Joly et al. (2016), McCulley et al. (2017), Ofstad et al. (2019), Blouin et al. (2021)
Roe deer (*Capreolus capreolus*)	Villerette et al. (2006a, b), Pagon et al. (2013)
Pronghorn (*Antilocapra americana*)	Kitchen (1974)
Impala (*Aepyceros melampus*)	Jarman (1979)
Kudu (*Tragelaphus strepsiceros*)	du Toit (1995)
Nayala (*Tragelaphus angasii*)	Kirby et al. (2008)
Wildebeest (*Connochaetes taurinus*)	Christianson et al. (2018)
Goitered gazelle (*Gazella subgutturosa*)	Blank et al. (2012)
Chiru (*Pantholops hodgsoni*)	Schaller and Junrang (1988), Schaller (1998)
Tibetan gazelle (*Procapra picticaudata*)	Jiang (2007)
Przewalski's gazelle (*Procapra przewalskii*)	Lei et al. (2006), Li et al. (2012)
Waterbuck (*Kobus ellipsiprymnus*)	Wirtz and Kaiser (1988)
Dall's and Stone's sheep (*Ovis dalli*)	Geist (1971), Corti and Shackleton (2002), Walker et al. (2006), Roffler et al. (2017)
Bighorn sheep (*Ovis canadensis*)	Geist (1971), Geist and Petocz (1977), Leslie and Douglas (1979), Morgantini and Hudson (1981), Shank (1982), Bleich et al. (1997, 2016), Ruckstuhl (1998, 1999), Fulbright et al. (2001), Tarango et al. (2002), Mooring et al. (2003), Mooring and Rominger (2004), Neuhaus and Ruckstuhl (2004), Brewer and Harveson (2007), Meldrum and Ruckstuhl (2009), Schroeder et al. (2010), Whiting et al. (2010), Hoglander et al. (2015), Donovan et al. (2021), Gastelum-Mendoza et al. (2021)
Mouflon (*Ovis musimon*)	Moncorps et al. (1997)

(continued)

Table 1. *(continued)*

Mouflon (*Ovis gmelini*)	Cransac et al. (1998), Le Pendu et al. (2000), Guilhem et al. (2006), Benoist et al. (2013), Marchand et al. (2015), Bourgoin et al. (2018)
Sardinian mouflon (*Ovis orientalis*)	Ciuti et al. (2008)
Merino sheep (*Ovis aries*)	Michelena et al. (2004, 2006, 2008), Gaudin et al. (2015)
Soay sheep (*Ovis aries*)	Pérez-Barbería and Gordon (1999), Pérez-Barbería et al. (2004, 2005, 2007, 2008), Ruckstuhl et al. (2006),
Tibetan argali (*Ovis ammon*)	Klich and Magomedov (2010), Singh et al. (2010b)
Darwin's wild sheep (*Ovis ammon*)	Li et al. (2017)
Marco Polo sheep (*Ovis ammon*)	Wang et al. (2019)
Corsican mouflon (*Ovis ammon*)	Bon and Campan (1989)
Blue sheep (*Pseudois schaeferi*)	De-Pin et al. (2013)
Mountain goat (*Oreamnos americanus*)	Holmes (1988), White (2006), Festa-Bianchet and Côté (2008)
Apennine chamois (*Rupicápra pyrenaica*)	Gerard and Richard-Hansen (1992), Pérez-Barbería et al. (1997), Unterthiner et al. (2012), Dalmau et al. (2013), Fattorini et al. (2019)
Chamois (*Rupicapra rupicapra*)	Shank (1985), Pérez-Barbería and Nores (1994), Ferretti et al. (2014), Corlatti et al. (2020)
Alpine ibex (*Capra nubiana*)	Francisci et al. (1985), Villaret and Bon (1995), Villaret et al. (1997), Bon et al. (2001), Neuhaus and Ruckstuhl (2002a), Grignolio et al. (2007, 2019), Ferrari et al. (2010), Tettamanti and Viblanc (2014)
Nubian ibex (*Capra nubiana*)	Gross et al. (1995)
Asiatic ibex (*Capra sibirica*)	Xu et al. (2012)
Siberian ibex (*Capra sibirica*)	Wang et al. (2008), Singh et al. (2010a), Han et al. (2019; 2020a, b; 2021)
Iberian ibex (*Capra pyrenaica*)	Alados and Escos (1987), Viana et al. (2018)
Markhor (*Capra falconeri*)	Ahmad et al. (2016, 2018)
Feral goat (*Capra hircus*)	Shi et al. (2003, 2005), Calhim et al. (2006), Dunbar and Shi (2008)
Muskox (*Ovibos moschatus*)	Oakes et al. (1992), Côté et al. (1997)
Steppe bison (*Bison priscus*)[extinct]	Guthrie (1990)
European bison (*Bison bonasus*)	Krasinska and Krasinski (1995), Ramos et al. (2016)
American bison (*Bison bison*)	Guthrie (1990), Berger and Cunningham (1994), Post et al. (2001), Mooring et al. (2005), Rosas et al. (2005), Schaad et al. (2005), Schuler et al. (2006), Kagima and Fairbanks (2013), Berger et al. (2014), Ranglack and du Toit (2015), Berini and Badgley (2017)
Yak (*Bos mutus*)	Berger et al. (2014)
Banteng (*Bos javanicus*)	Journeaux et al. (2018)
Cattle (*Bos taurus*)	Berteaux (1993), Lazo (1994)
Takin (*Budorcas taxicolor*)	Yan et al. (2016)
African buffalo (*Syncerus caffer*)	Sinclair (1977), Prins (1989), Halley and Mari (2004), Turner et al. (2005), Hay et al. (2008)
Giraffe (*Giraffa camelopardalis*)	du Toit (1990), Young and Isbell (1991), Ginnett and Demment (1997), du Toit and Yetman (2005), Mramba et al. (2017), Brown and Bolger (2020), Hart et al. (2021)
Camel (*Camelus dromedarius*)	Alkali et al. (2017)

species) likewise exhibit sexual segregation, but this phenomenon is best studied in mammals (~5,500 species). Even so, of the 27 orders of mammals (Fig. 3), only 8 orders have publications indicating the sexes segregate: Cetartiodactyla (artiodactyls, whales, and dolphins), Chiroptera (bats), Carnivora (seals and terrestrial carnivores), Primates (primates, including humans), Proboscidea (elephants), Diprotodontia (kangaroos and their relatives), Lagomorpha (hares, rabbits, and pikas); and Rodentia (an exceptionally large order with ~2,500 species of rodents). Evidence of sexual segregation in perissodactyls is lacking. Among terrestrial carnivores, canids (dogs and their relatives) are notably absent in the list of species that sexually segregate. Canids are one of the few mammalian families where there is considerable paternal investment in rearing young (Moehlman 1968, Kleiman and Malcolm 1981, Clutton-Brock 1991, Macdonald and Sillero-Zubiri 2004), which may reduce the propensity for sexual segregation. For ungulates, perissodactyls lack documentation of sexual segregation. Indeed, outcomes related to sexual segregation in mammals exhibit little overall relationship to their evolutionary histories or relatedness (Fig. 3). I caution, however, that sexual segregation may still be occurring in some taxa despite an absence of publications. Detecting sexual segregation can be difficult, especially if research is not designed specifically to do so (Bowyer 2004)—a point I address in more detail later.

This background sets the stage for better understanding sexual segregation in these exceptional large mammals. In the following chapters, I examine patterns of sexual segregation in ungulates, discuss its definitions, investigate hypotheses for its occurrence, explore the relevance of this phenomenon to theory and application, and clarify and organize results into a coherent theme that will inform future research on this fascinating subject. I will discuss how sexual segregation in ungulates concerns patterns of human behavior in the final chapter.

SUMMARY

The purpose of Chapter 1 is to provide a historical background on the topic of sexual segregation, review the ecology and behavior of ungulates to prepare for future discussions of sexual segregation, and to provide a basic understanding of the species of plants and animals that sexually segregate.

1. Charles Darwin first discussed the evolutionary underpinnings of sexual segregation and recognized this phenomenon occurred in an ungulate.

2. Ungulates are hooved mammals that make up the mammalian order Perissodactyla (odd-toed ungulates) and the terrestrial members (artiodactyls) of Cetartiodactyla (even-toed ungulates).

3. Ungulates have a near worldwide distribution and inhabit a diversity of environments that range across numerous habitats. They exhibit incredible variation in size, ranging from <10 to >2,000 pounds (907 kg).

4. Forest-dwelling ungulates tend to be small, monogamous, and lead a solitary existence, whereas those inhabiting open lands often are large, polygynous, and gregarious.

5. Ungulates consume a wide variety of plants and are categorized as browsers, intermediate feeders, and grazers, with body size typically extending from small to large along a browser to grazer continuum.

6. Artiodactyls are mostly ruminants, digesting (fermenting) plants in their rumen, whereas perissodactyls are hind-gut fermenters, with fermentation of plants occurring in their cecum and proximal colon.

7. Ungulates possess an impressive assortment of territorial and nonterritorial mating systems. Some species vary their mating system in response to environmental and social factors.

8. Ungulates have an array of life-history characteristics that differentiate them from small mammals; this slow-paced life history is

associated with strong density dependence, which is critical for understanding their population dynamics. Some ungulates lead a relatively sedentary existence, whereas others make extensive and spectacular long-distance migrations.

9. Whether large carnivores regulate populations of ungulates is context dependent. Circumstances when population regulation of ungulates by predation can occur include a predator pit, as well as under apparent competition. Predators are unlikely to regulate their own populations via territorial behavior.

10. Ungulates play a major role in affecting composition of plant communities and rates and pathways of nutrient cycling, depending on their population density.

11. An absence of predators may precipitate trophic cascades and cause ecological meltdowns at sufficiently high population density of ungulates.

12. Ungulates have been intertwined with human culture for over ~30,000 years; hunting and domestication of ungulates have played major roles in developing our civilization.

13. Historic declines of ungulates in North America and Europe occurred in the 1800s because of excessive harvest tied principally to economic incentives and commercial markets. Conservation efforts restored many of those populations. Today, wild populations of many ungulates are prospering, and legal hunting is allowed for many species.

14. Threats to the continued existence of ungulates occur worldwide and stem from a variety of causes, including illegal killing, habitat loss, and climate change. Even parks and other protected areas may not be sufficient to conserve ungulates in the face of an increasing human population.

15. Sexual segregation occurs in an array of plants and animals, yet its causes remain poorly understood and perhaps under reported. This phenomenon is best studied in mammals but has been described for only 8 of 27 mammalian orders. Among mammals, sexual segregation has been a focus for those interested in the ecology and behavior of ungulates. This background sets the stage for better understanding sexual segregation in these unique large mammals.

2 | Differences between the Sexes

Definitions of Sexual Segregation

How do differences between sexes lead to sexual segregation? Because of its widespread occurrence in ungulates, especially artiodactyls (see Table 1), I argue that the degree of sexual segregation meets the onerous conditions proposed by Williams (1966) for an *adaptation* (a change by which an organism or species becomes better suited to its environment) brought about via natural selection. Hypotheses that cannot explain the prevalence across species of ungulates are too narrow and will not suffice as a general explanation for sexual segregation (Bowyer 2004). Further, the common occurrence of sexual segregation among extant species of artiodactyls with dissimilar phylogenetic backgrounds infers causation by some general process rather than simply *phylogenetic inertia* (constraints on future evolutionary pathways that have been imposed by previous adaptations) (Bowyer and Kie 2004), although this process cannot be ruled out for perissodactyls, where sexual dimorphism in body size and sexual segregation have not been reported, even for well-studied species such as equids (horses, asses, and zebras) (Ransom and Kaczensky 2016). Indeed, sexual dimorphism among ungulates is a chief attribute of species exhibiting sexual segregation—

its occurrence, along with variation in sex ratios, spans millions of years (Berger et al. 2001a). Consequently, purely mechanistic or reductionist approaches that lack an evolutionary perspective are unlikely to provide an adequate explanation for sexual segregation (Bowyer 2004). An important question to help clarify the underpinnings of why the sexes segregate is, What is the adaptive significance of attributes underlying concepts proposed to explain sexual segregation? Undeniably, causes of sexual segregation in ungulates remain a hotly debated and contested topic in need of clarification. A critical detail is that most evidence indicates that all definitions for sexual segregation should relate ultimately to the differential use of space by the sexes. That point, then, should be the central factor connecting and refining definitions for sexual segregation, although there is a multitude of other opinions that I will evaluate and discuss.

Barboza and Bowyer (2000) observed that substantial confusion exists over specifically what constitutes sexual segregation. Indeed, the lack of a generally accepted operational definition for sexual segregation is the chief hinderance in describing and understanding this phenomenon and has remained so for >20 years. The traditional definition for sexual segregation in ungulates is the differential use of space or other resources by the sexes outside the mating season (rut) (Bowyer 2004). This definition emphasizes that space and related ecological components of *niche* (where an animal lives and how it makes its living) (Odum 1959) are major drivers of sexual segregation; this definition often invokes risk of predation and essential resources as primary causations. Importantly, definitions must be able to explain patterns of sexual aggregation and segregation on the landscape. This "ecological" segregation (sensu Mysterud 2000) results in use of space by the sexes being constrained to distinct areas at specific times. Several hypotheses for sexual segregation related to different morphological and associated physiological processes between the sexes likewise are linked with males and females using particular and relatively fixed features of the landscape when spatially separated—this will be considered in detail later.

An alternative view of sexual segregation in ungulates involves differences in activities of the sexes of sexually dimorphic species, which initially was conceived by Conradt (1998b) and Ruckstuhl (1998). The rationale and proposed mechanism for "social" segregation (Mysterud 2000) is based on higher fission rates of individuals in mixed-sex groups than in male-only or female-only groups, which results from asynchrony in activity patterns between the dimorphic sexes. Proponents of this definition contend that differences in activity provide a universal explanation for sexual segregation (Ruckstuhl and Neuhaus 2002, Conradt 2005, Ruckstuhl 2007). Indeed, this definition results in the differential use of space, but it gives no indication of where that space might be located (e.g., space use by males and females caused by differences in activity need not be constant and can overlap temporally); as well, this hypothesis does not offer an independent mechanism for seasonal changes in aggregation and segregation of the sexes. Nonetheless, this view of sexual segregation has gained wide acceptance.

An array of ideas concerning the role of social factors in affecting sexual segregation also have been proposed (I will deal with these individually under hypotheses for segregation of the sexes in Chapter 5 and 6). These notions include the need to assess fighting abilities of competitors, form dominance hierarchies, learn locations of water or places to give birth, and develop social affinities, and they include aggressive encounters, types of feeding behaviors, and avoiding competition with offspring, as well as others (Stewart et al. 2011). Again, most of these hypotheses are related to the differential use of space by the sexes but offer limited understanding as to where that space might occur. In the forthcoming section, I will review the underpinnings for sexual segregation and how they relate to the behavior and ecology of the sexes.

Sexual Differences in Morphology, Physiology, and Foraging Behavior

Sexual Dimorphism

Various forms of sexual dimorphism are prevalent among an array of vertebrate taxa (Short and Balaban 1994 for review). Darwin (1874) first observed that sexual dimorphism arose from natural selection operating differently on males and females because of their distinctive roles in reproduction, or from competition between the sexes for food or other resources. Darwin (1874) differentiated natural selection from selection relative to sex (i.e., sexual selection) (Andersson 1994). Those processes can function to produce or facilitate sexual dimorphism, and multiple causations may operate simultaneously (Hedrick and Temeles 1989). Two commonly hypothesized routes for the evolution of sexual dimorphism in vertebrates are competition within sexes for mates (sexual selection) and competition between the sexes for resources (natural selection). When species are polygynous, male-male combat can lead to enhanced reproductive success for winners and results in the evolution of increased body size of males (and their weapons) relative to the body size of females. Lande (1980) noted that a secondary process of natural selection for ecological displacement between the sexes might also occur. Fairbairn (1997), however, believed that such "ecological selection" would play only a subsidiary role in the evolution of sexual size dimorphism. Nonetheless, Slatkin (1984) suggested that ecological selection could result in sexual dimorphism even in the absence of sexual selection. Cassini (2017) proposed that for mammals in which females are larger than males fecundity played a role in enhancing size of females, a hypothesis first forwarded by Darwin (1874) and supported by research on bats (Myers 1978). Clearly, countervailing selection for increased size of females can reduce or reverse the degree of sexual size dimorphism (Bowyer et al. 2020a). In addition, Blanckenhorn (2005) proposed that searching for mates, as well as mechanisms of pre- and postcopulatory mate choice, should be considered within the process of sexual selection, but he cautioned that separating causes of sexual size dimorphism over evolutionary time from their present-day consequences for behavior and ecology could be challenging.

In the foregoing discussion, I used an example of males being larger than females in explaining sexual dimorphism. This pattern is characteristic of artiodactyls, but as I noted, does not occur for all vertebrates. Fairbairn (1997) provides a comprehensive overview of sexual dimorphism in plants and animals to help sort out such differences. For birds (Selander 1966, Storer 1966) and some reptiles (Shine 1989, Littleford-Colquhoun et al. 2019), females often are larger than males, although this pattern does not hold for some amphibians, including frogs (Nali et al. 2014). Care must be taken in implying causations across distantly related taxa. Indeed, Ralls (1977) cautioned that models for sexual dimorphism in birds would not suffice for mammals. Further, Soulsbury et al. (2014) reported a strong relationship between sexual size dimorphism and the strength of sexual selection in mammals but not in birds—the widespread presumption that sexual size dimorphism in birds results from sexual selection lacks

consistency and is more likely related to intersexual competition.

Proposed alternatives to sexual selection also exist for some highly sexually dimorphic mammals. Lindenfors et al. (2002) concluded that sexual size dimorphism was entirely the result of increases in size of males resulting from sexual selection among pinnipeds (seals, sea lions, walrus). Krüger et al. (2014), however, proposed that environmental factors promoted sexual dimorphism, which facilitated the evolution of polygyny in pinnipeds (i.e., characteristics resulting from natural selection lead to sexual selection). Although monomorphic pinnipeds tend not to segregate sexually (Staniland 2005), these large mammals may not be the best model for understanding sexual segregation in ungulates because of their vastly differing ecological and morphological characteristics, including carnivory and their adaptations to a mostly aquatic existence.

For mammals exhibiting sexual dimorphism in which males are larger than females, there is a clear pattern of *allometric* (disproportional) increases in degree of sexual dimorphism with increasing body size of males across species, a relationship known as *Rensch's Rule* (Fairbairn 1997, Sibly et al. 2012)—a pattern thought to be produced by sexual selection (Abouheif and Fairbairn 1997, Dale et al. 2007). This relationship is also evident for weaponry used in male-male combat (Kodric-Brown et al. 2006, Bro-Jørgensen 2007). No such association is apparent for species in which females are larger than males (Fairbairn 1997). Disproportional increases in sexual dimorphism with increasing body size are especially evident in bovids, cervids (deer), and macropods (kangaroos) (Sibly et al. 2012). Indeed, Polák and Frynta (2009) provide corroboration for Rensch's Rule for sheep, goats, and their relatives. Explanations for sexual segregation are most likely to be found among those diverse taxa. Too few extant species of elephants (African and Asian) are available to know whether Rensch's Rule might pertain, but a paleontological investigation of elephants might help clarify this issue. Nonetheless, those huge, sexually dimorphic mammals may provide a suitable ecological equivalent for ungulates that can inform our understanding of sexual segregation in some circumstances.

One aspect of sexual segregation in ungulates that is generally agreed upon is that this phenomenon is related to sexual dimorphism in body size (Mysterud 2000, Ruckstuhl and Neuhaus 2002, Bowyer 2004, Stewart et al. 2011). Ruckstuhl and Newhaus (2002) suggested that a body-weight difference of about 20% (males larger than females) in artiodactyls was the point after which sexual segregation became evident—such differences are most pronounced in larger species. Moreover, Loison et al. (1999) reported that small ruminants under 44–66 pounds (20–30 kg) did not exhibit sexual dimorphism, including small, monogamous species for which sexual segregation has not been reported (Dunbar and Dunbar 1980, Komers 1996). What, then, are thought to be the origins of sexual dimorphism, and how does this relate specifically to ungulates?

Perissodactyls exhibit limited sexual dimorphism, and sexual segregation is absent (Lenarz 1985, Neuhaus and Ruckstuhl 2002b). Nonetheless, equids and rhinos can be highly polygynous (Berger 1986, Rubenstein 1986, Duncan 1991, Rachlow et al. 1998), and male white rhinos (*Ceratotherium simum*) have larger horns than females (Rachlow and Berger 1997)—clearly, sexual dimorphism in body size for most of these ungulates is unrelated to degree of polygyny and subsequent male-male combat for mates. The evolution of sexual dimorphism in body size and weaponry occurs principally in those ungulates in which fighting between males for mating opportunities involves pushing, wresting, or ramming with horn-like structures (Geist 1966, Lundrigan 1996, Caro et al. 2003, Bro-Jørgensen 2007). Plainly, polygyny is not uniquely linked to sexual size dimorphism in all ungulates. When speed, agility, and aggressiveness are important in male-male combat, such as biting, slashing with canines, hooking with horns, or striking with hooves, increased size of a male may not be an advantage in competing with an opponent (Rughetti and Festa-Bianchet 2011, Bowyer et al. 2020a).

Some monomorphic artiodactyls are also polygynous or perhaps promiscuous, including the collared peccary (*Pecari tajacu*) (Bissonette 1982, Cooper et al. 2011) and vicuña (*Vicugna vicugna*) (Franklin 1983). Also, several species of oryx (*Oryx* spp.) do not exhibit strong sexual dimorphism in size, with both sexes possessing impressive rapier horns, yet the prevalence of mixed-sex groups indicates they are likely polygynous (Ruckstuhl and Neuhaus 2009, Robinson and Weckerly 2010). This lack of marked sexual dimorphism may be an instance when strong countervailing selection on females for large body size and horns reduced the degree of sexual dimorphism. Stankowitch and Caro (2009) demonstrated that, unlike males (Geist 1966), female bovids have larger horns that evolved principally as a defense against predators. Certainly, oryx are faced with an impressive collection of extant large African carnivores, such as African lions (*Panthera leo*) and leopards (*P. pardus*), which supports this observation. Obviously, selection for increased size in females may affect the degree of sexual dimorphism for some species (Myers 1978, Bowyer et al. 2020a).

Male cervids experience a swelling of their necks from enlarged muscles in preparation for strenuous male-male combat during rut (Lincoln 1971). A more elaborate and unique form of sexual dimorphism occurs in adult chamois. In those mountain ungulates, sexual dimorphism in skeletal size was about 5%; however, dimorphism in body mass was extremely seasonal. Males were about 40% heavier than females in autumn during rut, yet only 4% heavier in spring. Sexual dimorphism in chamois appears to result from greater summer accumulation of fat and muscles by males than by females—ostensibly to support rutting activities; male body mass declined rapidly during the rut (Rughetti and Festa-Bianchet 2011). Such sexual dimorphism in body size almost certainly is related to male-male combat during rut.

For artiodactyls, hypotheses for sexual dimorphism in size typically support the view that this phenomenon resulted from sexual selection (i.e., male-male combat for mates) (Jarman 1974, 1983; Weckerly 1998b; Loison et al. 1999; Pérez-Barbería et al. 2002) and not as a consequence of female choice per se. Indeed, the prevailing scenario for the evolution of sexual dimorphism in artiodactyls was proposed by Jarman (1974, 1983) using African antelopes (bovids) as a model. He suggested that ancestors of African antelopes dwelled in closed habitats (forests) and were unsocial, monogamous, and monomorphic. As grasslands flourished and forests diminished during the Miocene, forest-living antelopes spread into the plains. Janis (1982) provides paleoecological support for this change. Jaeggi et al. (2020) offer evidence that social ungulates evolved from more solitary ones. Those plains-dwelling antelopes began developing adaptations, including increased sociality related to foraging patterns and predation, that allowed them to persist in open habitats. Takada and Minami (2021) reported that solitary Japanese serow (*Capricornis crispus*) increased the size of social groups as they occupied more open habitat. Large social groups offered the prospect for

males to monopolize mating opportunities and promoted the evolution of polygyny. The advent of polygynous mating, and associated male-male combat, led to selection for large male body size and, consequently, sexual dimorphism and the elaboration of horn-like structures used for varied modes of fighting (Geist 1966, Lundrigan 1996, Caro et al. 2003, Bro-Jørgensen 2007, Plard et al. 2011), possibly including selection for enhanced fighting skills (Pérez-Barbería and Yearsley 2010). Males with the large horn-like structures tend to have high mating success (Bowyer 1986a, Kruuk et al. 2002, Vampé et al. 2007, Markussen et al. 2019). Pérez-Barbería et al. (2002) offer phylogenetic support for the pattern of evolution proposed by Jarman (1974, 1983). This general model has promise for understanding the progression in which open habitat, associated increases in group size, and subsequent polygynous mating resulted in the evolution of sexual size dimorphism among open-land ungulates (Bowyer et al. 2020a)—and ultimately how this sequence of events leads to sexual segregation (Fig. 4). Strong feedbacks exist from sexual segregation on an absence of paternal

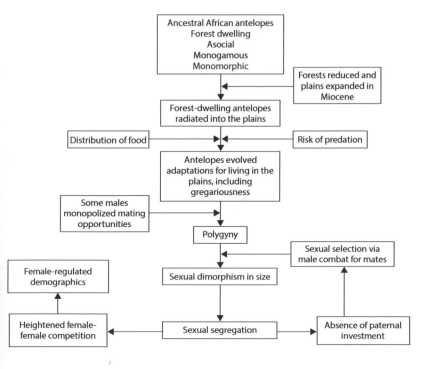

Fig. 4. A pattern of evolution for African antelopes (Bovidae) explaining links among habitat, sociality, degree of polygyny, and sexual dimorphism and explaining how these environmental and social factors lead to sexual segregation. (Adapted from Jarman 1974, 1983; modified from Bowyer et al. 2020a)

investment in young and, ultimately, sexual dimorphism in artiodactyls (Fig. 4). Likewise, effects of sexual segregation on female-female competition have strong effects on population dynamics of ungulates, which will be discussed in detail in Chapter 8.

Cassini (2020a, b) suggested that a model of sexual dimorphism based on natural selection, similar to that proposed for pinnipeds, might exist for artiodactyls, wherein sexual dimorphism was a cause rather than a result of polygyny. Indeed, Cassini (2022) proposed that sexual dimorphism in artiodactyls was a function of intersexual competition (natural selection) rather than male-male combat (sexual selection) and that sexual segregation resulted from that evolutionary pathway. The difficulty with this view of evolution is that experimental evidence demonstrates that increasing population density rather than intensifying competition between the sexes results in greater niche partitioning and the avoidance of competition (Kie and Bowyer 1999, Stewart et al. 2015; Chapter 7). Moreover, increasing population size decreased the amount of sexual segregation. Thus, there is no evolutionary mechanism to sustain sexual dimorphism in artiodactyls if intersexual competition is the cause, because the sexes partition niche space. The best explanation for the evolution of sexual segregation is the pathway illustrated in Figure 4.

Cassini (2020a, b) correctly noted that molecular data for assigning paternity was more resolute than methods using mating group size as an index to polygyny, and that multiple reproductive tactics by males might affect the evolution of sexual size dimorphism. Both metrics, however, reflect degree of polygyny and do not substantially alter the role of sexual selection in sexual dimorphism. In Chapter 1, I noted that many ungulate mating systems appear fixed, but alternative mating tactics exist for some species, and that those behaviors could result in flexible mating systems (Bowyer et al. 2002a).

Variation in reproductive success for most polygynous artiodactyls occurs because large, dominant males sire more offspring than their smaller subordinates (Bergeron et al. 2010, Willisch et al. 2012). Moreover, overall lifetime reproductive success is higher for males in more sexually dimorphic species, such as red deer, compared with less-dimorphic species, such as roe deer (*Capreolus capreolus*) (Vampé et al. 2008). Alternate mating tactics, however, exist. Such tactics are thought to be condition (i.e., state) dependent. Individuals may alter their mating tactics based on internal factors, such as age or size (Corlatti et al. 2020), external features of the environment, including distribution of resources (Bowyer et al. 2020a, Corlatti et al. 2020), or social factors (Isvaran 2005). So long as the same individual engages in alternate mating tactics over time, such as taking risks in attempting to sneak copulations when younger and exerting dominance via aggression when they are older and large enough to do so (Bowyer 1986a), there should be limited changes in selection for size dimorphism in males. Indeed, Thalmann et al. (2015) reported that for black-tailed deer strong relationships existed between age and size of antlers, as well as body mass and antler size (Fig. 5), although density dependence added variance to those regressions. Male body size continues to increase long after females have reached asymptotic body growth in dimorphic ungulates (Reimers et al. 1983, Schwartz et al. 1987). Further, Stewart et al. (2000) demonstrated that male moose do not invest fully in horn-like structures until body growth is complete (Fig. 6), which might allow males to follow multiple strategies throughout their lives. Moreover, moose and reindeer may alter their mating system within a single rut by adjusting their behavior to the availability of estrous females (Bowyer et al. 2011, Weladji et al. 2017). When ecological factors come into play, however, the possibility exists for multiple mating tactics to operate in the same population with the potential for both small and large individuals to garner copulations (Cassini 2020b, Corlatti et al. 2020).

For polygynous bovids inhabiting mountainous terrain, mating opportunities exist for both subadult and adult males. Such species exhibit slow body

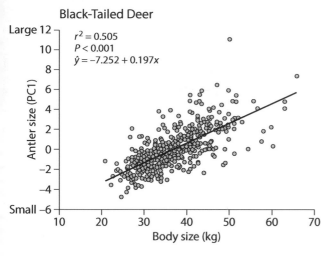

Fig. 5. Regression of antler size (*PC1*) on body mass of harvested male black-tailed deer 1979–1997 from central California. (Modified from Thalmann et al. 2015)

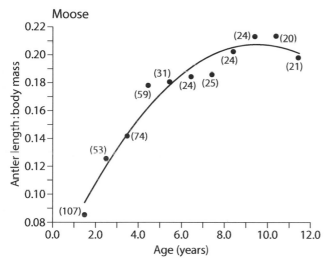

Fig. 6. Relationship between the ratio of antler length (cm) to body mass (kg) compared with age in years for Eurasian moose. Regression model: $Y = 0.04 + 0.3x - 0.002x^2$; $R^2 = 0.961$, $P < 0.0001$, $n = 11$ weighted by sample sizes (*n*) shown on figure. Data on age, antler length, and body mass were from Prieditis (1979). Note that full investment in antler size does not occur until moose are fully mature. (Modified from Stewart et al. 2000; originally published in *Alces*)

growth, and young males are lighter and more agile than older ones. The rugged terrain allows subordinate subadults to evade adults and impedes dominant adults from preventing access of young males to some females (Bowyer et al. 2020a). This rugged environment selects for mobility, as typified by alpine ibex (*Capra nubiana*) (Apollonio et al. 2013). Indeed, for bighorn sheep (Coltman et al. 2002) and blue sheep (*Pseudois nayaur*) (Lovari and Ale 2001), subadult males can sire a considerable proportion of neonates, but usually less than older, dominant males. Bowyer et al. (2020a) noted that, in contrast to mountain ungulates, fast-growing cervids

and bovids that inhabit less-rugged terrain include large adult males, which can monopolize females and thereby limit mating opportunities by excluding most young males from access to potential mates (Pemberton et al. 1992, Wilson et al. 2011, Ciuti and Apollonio 2016).

I do not believe that multiple mating tactics would have a major effect on sexual dimorphism for most artiodactyls, and when they do so, would involve weakly sexually dimorphic species. If this pattern was prevalent, strong relationships with dimorphism in keeping with Rensch's Rule would not occur. These are critical details because sexual segregation is convincingly related to sexual size dimorphism in artiodactyls. Further, the sequence of evolution for sexual dimorphism and sexual segregation delineated in Figure 4 is most congruent with the ecology and behavior of artiodactyls. Hence, sexual dimorphism is mostly a product of sexual selection, but there may be strong ancillary feedbacks from natural selection that reinforce that process, especially with respect to differences in morphology between the sexes.

Teeth Morphology and Diet

I noted in Chapter 1 that marked differences in the dental morphology of ungulates were related to diet, especially differences between those eating browse and grass. A proposed avenue for sexual segregation was that competition between the sexes for forage, driven by differences in feeding behavior, was an outcome of variation in dental morphology between sexes (Clutton-Brock et al. 1987; Illius and Gordon 1987; Main and Coblentz 1990, 1996). Pérez-Barbería and Gordon (1998, 2000), however, reported that incisor arcades were altered in an isometric (proportional) fashion with body mass of artiodactyls (i.e., those characteristics did not offer a reasonable explanation for sexual dimorphism in size). Loison et al. (1999) also noted that differences in feeding style (browsing, intermediate feeding, or grazing) were not associated with sexual size dimorphism in ungulates.

For the moose, a large browser, incisor breadth differed little between males and females (males ~6% broader), even for a species with huge differences in body mass between sexes (males 40% larger) (Spaeth et al. 2001; Fig. 7). Males continued to gain body mass long after maximal incisor breadth was attained, but maximal body mass and incisor breadth were reached at a similar age in females (Fig. 7). Incisor breadth of moose scaled isometrically (proportionally) with a projected body mass to $kg^{0.23}$, which was considerably less than estimates for a variety of other ungulates—$kg^{0.33}$ (Clutton-Brock and Harvey

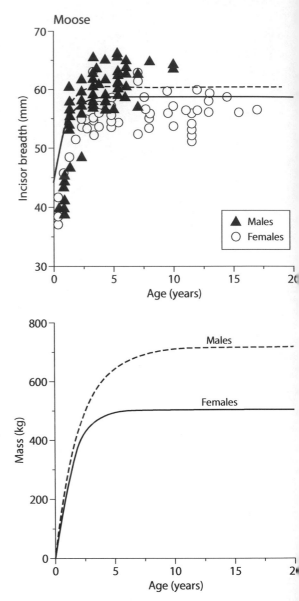

Fig. 7. Relationship between breadth of the lower incisor arcade for 98 male (*dashed line*) and 88 female (*solid line*) Alaska moose (*above*), and the relation between body mass and age (*below*). Von Bertalanffy equations for incisor breadth and age are provided in Spaeth et al. (2001). The relation between body mass and age was modified from Schwartz et al. (1987). Note that the size of the incisor arcade asymptotes at 4 years for males and females, but that female body size asymptotes a much earlier age than for males. (Modified from Spaeth et al. 2001; originally published in *Alces*)

1983) and kg$^{0.40}$ (Gordon and Illius 1988). Caution should be used in extending comparisons from interspecific relationships because those predictions may not hold for intraspecific ones (Barboza and Bowyer 2000). Incisor arcades were little altered with respect to sexual size dimorphism in moose. Weckerly (1993) reported similar results for black-tailed deer. Clearly, mouth morphology is unrelated to degree of sexual dimorphism and thereby fails to offer a reasonable explanation for sexual segregation in ungulates. Moreover, dietary differences occur seasonally for ungulates inhabiting temperate and arctic environments, but mouth architecture of adults remains relatively constant. Thus, there is a lack of morphological variation in mouth architecture associated with seasonal variation in the degree of sexual segregation. The dental array of ungulates is best explained as an outcome from animals feeding on diverse diets rather than its cause.

Brashares et al. (2000) noted that African antelopes with a more-selective diet tended to occur in smaller groups and had lower body weights than species with unselective diets, factors that hold potential to affect group living and thereby sexual segregation. An exceptionally large cervid (moose; Fig. 7), however, exhibits high diet selectivity while browsing extensively on willows (*Salix* spp.) (Bowyer and Bowyer 1997, Weixelman et al. 1998, Bowyer and Neville 2003), which indicates that the proposed pattern may not be generally applicable to ungulates. I further suggest that diet selectivity is an outcome rather than a cause of inhabiting closed versus more open lands, which also influences group size and body weight (topics I will revisit later).

Gastrointestinal Characteristics and Related Physiology of Ruminants

Numerous researchers have examined the relevance of the relationship of metabolism and gut capacity to body size as an explanation for diet composition in ruminants (Short 1963, Prins and Geelen 1971, Sinclair 1977, Hanley 1980, Demment 1982, Demment

and Van Soest 1985). Indeed, variation in body size of ruminant species relative to their gut capacity and metabolic rate has been offered as an explanation for the diversification of those herbivores into a variety of foraging niches (termed the *Bell-Jarman Principle*). Across species, a strong relationship has been reported between gut capacity and body weight that is isometric (proportional, with a scalar equal to 1), which is evident over an array of feeding styles (Demment 1982, Bunnell and Gillingham 1985, Demment and Van Soest 1985, Clauss et al. 2007, Ramzinski and Weckerly 2007). The relationship between metabolic rate and body weight, however, is thought to be allometric (disproportional), with scalers between 0.67 and 0.75 (Bunnell and Gillingham 1985). That ratio (the isometric scaler of gut capacity to the allometric scaler of metabolic rate) has been invoked to explain differences in digestive efficiency among species of ruminants. Larger species have a greater gut capacity (in the rumen and reticulum), which allows for greater food intake, longer fermentation times (slower rates of food passage), and the ability to digest poorer quality forages than smaller species (Demment and Van Soest 1985). The Bell-Jarman Principle proposes that the digestive consequences of that ratio explain forage partitioning by species across the substantial scope of sizes for ruminants (Bell 1970, Jarman 1974).

Pérez-Barbería et al. (2008) proposed that the Bell-Jarman Principle might also be applied to sexually dimorphic ungulates to help explain sexual segregation among those large herbivores. A primary assumption for the Bell-Jarman Principle to offer a viable explanation for sexual segregation is that this process, which operates among species, also will be manifest within a species. Barboza and Bowyer (2000, 2001) cautioned that this might not be the case. Indeed, Jenks et al. (1994) demonstrated greater mass of ruminal fill for female than for male white-tailed deer. Moreover, Weckerly (2010) reported that the intraspecific rumen-reticulum capacity scaled allometrically with body size in white-tailed deer. Measurement of rumen fill can be affected by timing of sampling and dietary differences between sexes. Nonetheless, if correct, the scalar for the rumen-reticulum relationship was not greater than the scalar for the metabolic relationship, which lessens the likelihood that the Bell-Jarman Principle will explain dietary differences between the sexes (Weckerly 2010). Importantly, the existing hypotheses for niche partitioning among species of large herbivores are not sufficient to explain sexual segregation, because those ideas will not account for temporal patterns of segregation and aggregation between sexes (Bleich et al. 1997, Barboza and Bowyer 2000).

Subsequently, a new hypothesis was formulated to explain sexual segregation based on an allometric model of metabolic requirements, minimal food quality, and digestive retention—the *gastrocentric hypothesis* (Fig. 8). This model predicts that male deer consume abundant forages high in fiber because the large rumen-reticulum prolongs retention (thereby increasing digestion) and permits greater use of fiber for energy than in nonpregnant or nonlactating females. Low density of animals, high abundance of food, and adaptations of rumen microbes keep large males on fibrous forages until quantity of food declines (Barboza and Bowyer 2000). Indeed, males and females possess different gastrointestinal microbes when sexually segregated (Zhu et al. 2020). Smaller-bodied females, in comparison, are more proficient at postruminal digestion of food, especially when forage intake increases concurrently with requirements for energy and protein during

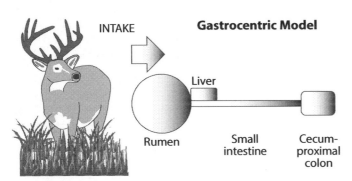

Gastrocentric Model

Large male — high gut capacity & long retention

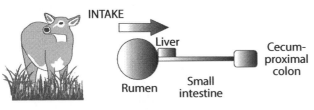

Nonreproductive female — base plan

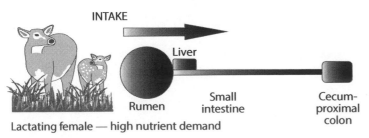

Lactating female — high nutrient demand

Fig. 8. Model of intake and digestive function in nonreproductive females (*middle*) compared with large males (*above*) and lactating females (*below*). Width of arrows reflects amount of food intake, length of arrows indicates rate of digesta passage, and shading indicates density of nutrients in food (darker = higher quality). Diagrams of the digestive tract are shaded to reflect potential changes in fiber content of food (darker = greater density of nutrients in food) for males and increases in postruminal size (small intestine and cecum-proximal colon) and function of lactating females. (Adapted from Barboza and Bowyer 2000, Zimmerman et al. 2006, Stewart et al. 2011; from Bowyer et al. 2020a)

reproduction. High demands in females for absorption of nutrients during lactation and growth stimulate increases in intestinal and liver tissue (Barboza and Bowyer 2000), as well as stimulating an increase in rumen size and number of papillae to better absorb nutrients (Zimmerman et al. 2006), thereby increasing the cost of maintenance and reinforcing differential use of habitats or forages when sexes are segregated (Fig. 8). Moreover, demands for energy and protein during lactation increase dietary requirements for pregnant females above levels required for nonpregnant females and for males. Consequently, differences in feeding activity between the sexes are likely a consequence of metabolic demands rather than the cause (Barboza and Bowyer 2001). Moreover, the Bell-Jarman Principle should not be conflated with the gastrocentric hypothesis, which includes differences in digestive morphology and physiology for females not included the Bell-Jarman Principle. Differences in digestive morphology provide a reason for females segregating from males around the time of parturition, and then as their digestive tract changes, aggregating with males during rut. The Bell-Jarman Principle lacks a mechanism to explain sexual aggregation. Moreover, females reduce forage intake as they attain "target" body mass during rut, although small females may continue to forage. Large (but not small) males may fast. Consequently, forage intake and demands of the sexes are heterogenous. Metabolic demand cannot explain sexual segregation because no tradeoff exists between the proximate factor (food) and the ultimate factor (sex) (Thompson and Barboza 2017).

The gastrocentric hypothesis also offers a reason for males not using areas of high quality occupied by females around the time of parturition. Males possess a larger ruminal volume that can accommodate digestion of coarse plant fibers, whereas females on a similar diet may require more processing by chewing and rumination (Gross et al. 1995, Barboza and Bowyer 2000). Slow rates of passage and a fibrous diet favor cellulolytic microbes in the rumen of males. Conversely, females have a portion of rumen microbes better adapted to faster passage of food, and they acquire a greater proportion of nutrients from cellular contents than from cell walls of plants (Barboza and Bowyer 2000). Males would be prevented from quickly switching between diets of differing quality because such changes would disrupt ruminal fermentation and risk excess production of gases and bloat, malabsorption, and scouring (Van Soest 1994, Gordon and Illius 1996, Barboza et al. 2009). Males can obtain adequate nutrition on lower-quality, more abundant foods than those high-quality foods required by pregnant and lactating females (Barboza and Bowyer 2000), which provides a plausible explanation for sexual segregation.

The Role of Resources and Predation Risk in Gregariousness

Resources and Topography

Distribution and quality of resources can modify gregariousness in ungulates (Jarman 1974, Fryxell 1991, Street et al. 2013). Jarman (1974) hypothesized that the dispersion of food items was a key element influencing the degree of sociality in large herbivores. Ungulates that are solitary or live in small groups often inhabit woodlands or forests, selectively foraging on dispersed leaves, stems, and mast of browse (woody vegetation) or eating herbaceous vegetation (forbs). Thus, Jarman (1974) proposed that in coarse-grained habitats with a patchy distribution of food items, such as woodlands, foraging by one animal limits food availability to others by removing the whole food item (herb, stem, or leaf). The ensuing dispersion of food items would result in individuals avoiding areas where others had foraged previously, causing a wide distribution of animals and impeding group formation.

Ungulates dwelling in open grasslands and savannahs typically have large body sizes and occur in large groups. Those ungulates exhibit low diet selectivity, feeding principally upon more evenly distributed grasses. In such fine-grained habitats, where food items are more evenly distributed, ungulates remove foods a little at a time. Forage is reduced, but the distribution of food items remains relatively uniform. Accordingly, individuals can feed closer together and potentially form groups (Jarman 1974). Those overall patterns for the distribution of forage allow large herbivores to alter group size but fail to explain *why* they should do so (Bowyer et al. 2001a). Street et al. (2013) concluded that forage-mediated aggregations did not offer a stand-alone explanation for gregariousness in ungulates. A tendency exists for group size to increase with population density among species of ungulates (Putman and Flueck 2011), but exceptions to this general pattern can exist within a species (Kie and Bowyer 1999), indicating the importance of other factors in influencing the size of groups.

Herbivore optimization can result from groups of large herbivores increasing productivity of plants and rates of nutrient cycling in places where they have foraged and deposited urine and feces previously (McNaughton 1979a, Molvar et al. 1993, Stewart et al. 2006, Guernsey et al. 2015). Ungulates may return to such areas to acquire higher-quality foods, which would further foster sociality. Clumped resources also can affect size of social groups. Bowyer et al. (2001a), however, reported that the size of social groups of mule deer was unrelated to the size of resource patches—resources set limits on the number of mule deer available to form groups but failed to adequately explain group size for those large herbivores.

The quality and distribution of resources can affect the behavior of ungulates (Belovsky 1981, Fryxell 1991), although effects of scale in relation to the distribution of resources are important (Bowyer et al. 2001a, Bowyer and Kie 2006). For animals to be gregarious, sufficient resources must exist to allow group living (McNab 1963). Moreover, some threshold for a specific resource may exist at which group formation occurs (Schoener 1968). Further, necessary resources help set the ecological carrying capacity (K), thereby determining the number of

large herbivores that a specific area can support (McCullough 1979, Boyce 1989, Bowyer et al. 2014). Similarly, heterogeneity of the landscape can determine the size and arrangement of home ranges for large herbivores (Kie et al. 2002), which influences the number of animals that can associate to form groups. Indeed, Jhala and Isvaran (2016) demonstrated that a decline in group size of blackbuck corresponded with increasing patchiness of habitat. Clearly, variation in distribution of resources can influence the size and spacing of social groups, and thereby holds potential to affect sexual segregation.

A suite of life-history characteristics, morphology, and behaviors have been associated with ungulates occupying open as opposed to closed habitats, including differences in group size (Estes 1974, Peek et al. 1974, Leuthold 1977, Gosling 1986, Geist and Bayer 1988, Bowyer 2020a). Indeed, Bowyer et al. (2002a) reported that Alaskan moose from areas with open tundra had large antlers than those inhabiting closed taiga forest, although nutrition also was involved. Unaccountably, Loison et al. (1999) failed to detect effects of habitat structure on degree of sexual dimorphism in size for ungulates. The authors offered several plausible explanations for that outcome, but I believe there was another potential cause. Mountain ungulates typically inhabit open habitat but often occur in steep and rugged terrain that limits gregariousness (Bowyer et al. 2020a). This is an extreme example, but terrain brokenness and steepness have strong effects on limiting the size of social groups, even for species inhabiting less-rugged landscapes (Bowyer et al. 2001a). Nearly a 5-fold decrease in the maximum size of groups occurred with increasing steepness or brokenness of topography for mule deer (Fig. 9). Terrain may affect the ability of group members to maintain contact with one another, detect predators, clump together when confronted by a predator, or effectively elude pursuit (Bowyer et al. 2001a). Plainly, a strong potential exists for variation in group size to result from features in the landscapes in which the groups occur, even for species occupying open habitats.

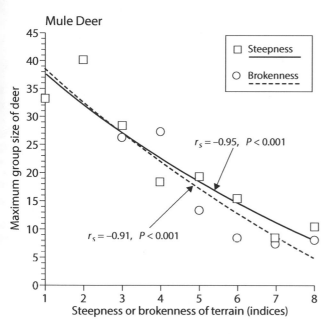

Mule Deer

□ Steepness
○ Brokenness

$r_s = -0.95$, $P < 0.001$

$r_s = -0.91$, $P < 0.001$

Maximum group size of deer

Steepness or brokenness of terrain (indices)

Fig. 9. Relationships between maximum size of mule deer groups and steepness or brokenness of terrain in the Cuyamaca Mountains of southern California. Categories of slope steepness range from 0% to 14% in increments of 1.75%; brokenness ranges from 0 to 10 m in depth of gullies and washes running parallel to slope exposures in increments of 1.25 m. (From Bowyer 2001a; originally published in *Alces*)

Predation

Risk of predation can play a major role in sexual segregation (Bleich et al. 1997, Bowyer 2004, Ciuti et al. 2004). Considering predation risk without bearing in mind gregariousness is largely pointless, however, because degree of sociality is an integral part of ungulates thwarting predators (Bowyer et al. 2020a), and it is essential to understanding the evolution of sexual segregation (Fig. 4).

I do not propose to examine all potential costs and benefits of group living in vertebrates; these have been enumerated previously (Alexander et al. 1974, Krause and Ruxton 2002). Nonetheless, when ungulates employ antipredator behavior—including the use of concealment cover (Hirth 1977, Bowyer et al. 2001a), vigilance, flight, escape terrain (Bleich 1999), or group formation—the probability of detection or capture can be reduced. Sometimes this behavior may involve an active defense against predators (Kruuk 1972, Molvar and Bowyer 1994, Bleich 1999, Caro 2005). Females with young were more vigilant than either females without young or males in Tibetan argali (*Ovis ammon*), ostensibly a response to predation risk from wolves (Singh et al.

2010a). Indeed, risk of predation is thought to influence gregariousness of ungulates through benefits of increased group size in open-land species, which ensue because of more eyes, ears, and noses with which to detect predators at distances where successful pursuits are less likely (Roberts 1996). Costs related to competition also may occur for ungulates in large groups (Uccheddu et al. 2015).

When a predator (or hunting group) can catch only a single animal at a time, which often is the case for ungulates and the large carnivores that prey upon them, additional advantages may arise from social groupings. A lone animal has a greater "domain of danger" than individuals in a group and, consequently, a higher probability of being selected as prey than animals occurring in a *selfish herd*, an outcome termed *dilution effects*. Those selfish individuals accrue benefits from grouping related to the self-interest of increased survival via a lowered probability of attack, which need not be related to cooperation or kin selection (Hamilton 1971). Morton et al. (1994) noted that individuals moving toward their nearest neighbor provided an additional antipredator tactic for the selfish herd. Dehn (1990) reported potential benefits from both dilution effects and vigilance for ungulates occurring in large groups. Indeed, multiple benefits may accrue to open-land ungulates that live in social groups, many of which are not mutually exclusive (Bowyer et al. 2001a).

Substantial benefits also accrue to forest-dwelling species from living in small groups (Hirth and Mc-Cullough 1977, Putman 1988). Noise and odors from large groups of ungulates moving through dense vegetation might interfere with detection of ambush or stalking carnivores, whereas small groups can be more secretive (Bowyer et al. 2001a). Exceptions to this behavior, however, exist. White-lipped peccaries (*Tayassu pecari*) can form exceptionally large groups in densely vegetated areas (Reyna-Hurtago et al. 2009).

A clear pattern is evident for open-land ungulates, such as American bison, to increase group size as they move farther from the forest's edge (Bowyer et al. 2007; Fig. 10), a pattern also reported for black-buck (Jhala and Isvaran 2016). Similarly, mule deer increase the size of social groups with greater distance from concealment cover, with males venturing farther from cover than either females or young (Bowyer et al 2001a). Indeed, mule deer alter group size with respect to the cover provided by the habitats in which they occur (Fig. 11), an outcome also

demonstrated for white-tailed deer (Hirth 1977). Concealment cover, visibility, and proximity to a refuge have important effects upon the perception of predation risk in mammals (Camp et al. 2012, Hefty and Stewart 2018). Larger groups of mule deer in meadows spent more time feeding and less time in alert-alarm postures than smaller groups; foraging efficiency (percent of active time spent foraging) (Berger 1978) increased with size of groups. Aggressive interactions also occurred more often in large groups, but per capita rate of aggression declined with increasing group size (Bowyer et al. 2001a).

Antipredator behaviors, then, likely promoted other social behaviors related to foraging efficiency that favored group living in open-land ungulates. Molvar and Bowyer (1994), however, demonstrated that moose formed social groups in response to predation risk without the concomitant benefits of enhanced foraging efficiency, which indicates that some advantages of group living may be secondarily evolved. Molvar and Bowyer (1994) also documented that selective foraging declined as moose moved from taiga into open tundra; moose took larger and less nutritious bites of willow as they moved farther from cover (Fig. 12). Weixelman et al. (1998) similarly reported that moose altered their selection (use/availability) for common species of browse as they

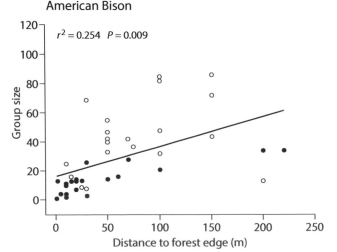

Fig. 10.   Linear relationship ($\hat{Y} = 16.95 + 0.19x$) between group size and distance to the forest edge for all groups of American bison with large males (*open circles*) and without large males (*closed circles*) during summer observed on the Delta Junction Bison Range in interior Alaska. (From Bowyer et al. 2007)

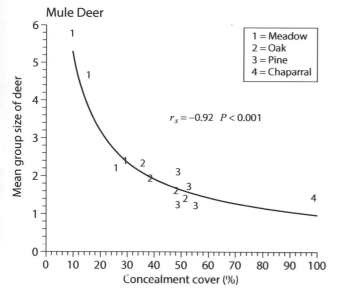

Mule Deer

1 = Meadow
2 = Oak
3 = Pine
4 = Chaparral

$r_s = -0.92$ $P < 0.001$

Fig. 11.   Relationship between mean group size of mule deer and concealment cover for 4 major plant communities measured with a density board during each season in the Cuyamaca Mountains of southern California. Chaparral (*4*) was 100% cover during all seasons. (From Bowyer 2001a; originally published in *Alces*)

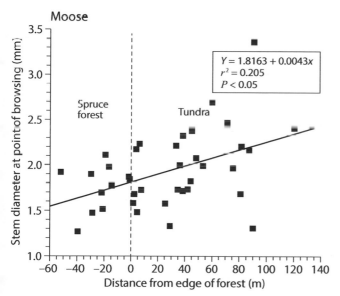

Moose

$$Y = 1.8163 + 0.0043x$$
$$r^2 = 0.205$$
$$P < 0.05$$

Spruce forest

Tundra

Fig. 12.   Relationship between twig diameter at point of browsing (*solid line*) by moose on stems of willow (*black squares*) and distance from the edge of the forest (*dashed line*) in summer, interior Alaska. Negative values for distance indicate that moose were foraging within the forest. (From Molvar and Bowyer 1994)

moved farther from cover, becoming less selective at greater distances from the edge of burns, which provided little concealment cover. Edwards (1983) likewise noted that moose shifted their diet as they attempted to avoid predators. Ciuti et al. (2004), however, demonstrated that female fallow deer would accept greater predation risk to obtain higher-quality forage.

Group size in bighorn sheep increased as females ventured farther from escape terrain (Berger 1991).

Female Dall's sheep likewise increased group size as they moved farther from rugged terrain (Rachlow and Bowyer 1998). Schroeder et al. (2010) also documented that female bighorn sheep occurred closer to escape terrain but were in larger groups than males; there were few differences, however, in foraging efficiency or vigilance behavior between the sexes (Fig. 13). Hence, the sexes employed differing antipredator tactics but achieved similar outcomes with respect to foraging behavior. Vulnerability to

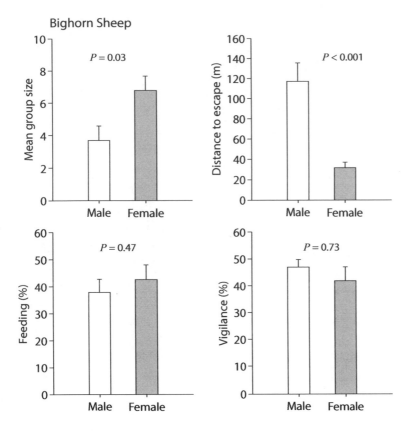

Bighorn Sheep

Fig. 13. Mean group size (+SE) for distance to escape terrain (*above*) and percent of time spent feeding and being vigilant (*below*) for 15 male and 14 female bighorn sheep during winter in the Sierra Nevada, California. (From Schroeder et al. 2010)

predation markedly affected the behavior of the sexes in these highly dimorphic ungulates. Such differences likely were related to the smaller size of females, ostensibly making them more vulnerable to predation (including to a wider array of potential predators) than larger males, as well as increasing susceptibility of their young to predation (Clutton-Brock et al. 1982, Li et al. 2009). Indeed, mountain ungulates often select steep and rugged terrain for birthing of young (Bleich et al. 1997, Rachlow and Bowyer 1998, Han et al. 2021), evidently an adaptation to help thwart predators.

Other behaviors by ungulates to hinder predators also exist. Parturient caribou may disperse from large carnivores (Bergerud et al. 1984, Bergerud and Page 1987). Barten et al. (2001) documented that parturient female caribou moved to high-elevation sites that were mostly above the distribution of gray wolves and grizzly bears, but females without young occurred at much lower elevations (Fig. 14). More-

over, if a female lost her neonate, she moved to lower elevations. Female bighorn sheep with young occurred on steeper slopes and in more rugged terrain than adult males or females without young (Bleich et al. 1997). Berger et al. (2014) reported that female wild yaks (*Bos mutus*) occurred in wetter or more rugged topography than males. Moreover, female yaks with young used more undulating or steeper slopes than females without young. An experiment by Sarmento and Berger (2020) revealed increased use of extremely steep slopes by mountain goats (*Oreamnos americanus*) following exposure to an imitation grizzly bear, indicating that such precipitous terrain was a critical component of habitat for those mountain ungulates when faced with a potent predator. Predation risk experienced by the sexes is an important component of their habitat selection for steep and rugged terrain, especially for alpine ungulates.

Gregariousness comprises more than just an adaptive response to changes in habitat structure or to-

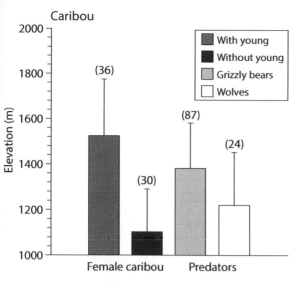

Fig. 14. Median (+ 1/2 interquartile distance) elevation at which female caribou with and without young and predators occurred during peak parturition, Wrangell–St. Elias National Park and Preserve, Alaska. Sample sizes are in parentheses. (From Barten et al. 2001)

pography. Ungulates likely communicate information concerning the presence of a predator to other group members through alarm behaviors, including distinctive pelage markings, piloerection, specialized gaits, alarm vocalizations, pheromones, or some combination thereof (Hirth and McCullough 1977; Caro 1986; Bowyer et al. 1991, 2001a). Promoting group cohesion as an antipredator behavior offers the most likely explanation for such displays (Hirth and McCullough 1977, Bowyer et al. 1991). Other potential explanations, however, have been forwarded to explain the evolution of rump-patch displays.

Estes and Goddard (1967) noted that individuals often engaged in alarm signaling with rump-patch displays when confronted with, or fleeing from, a predator. Those authors hypothesized that such displays served to warn other members of the group of potential danger. Such signaling, however, likely makes the individual more obvious to predators, and consequently increases the risk to the signaler (Maynard Smith 1965). Thus, the evolution of rump-patch displays would be difficult to explain unless kin selection, group selection, or reciprocity were in-

voked. This outcome is unlikely because many groups of ungulates are neither stable nor always composed of close relatives, which would be necessary for kin selection or reciprocity to evolve (Bowyer et al. 1991), leaving group selection as the other unlikely alternative (Williams 1966).

Guthrie (1971) proposed that rump-patch displays evolved initially as an intraspecific submissive gesture to appease dominant individuals. Presentation of the rump patch by subordinates serves to ameliorate aggressive behavior by dominants. Geist (1971) described the mimicking of a copulatory posture to appease dominant individuals in mountain sheep. Thus, rump patches ultimately acquired the additional purpose of a warning mechanism by prey presenting a submissive display to predators, but whether this hypothesis will explain rump patches in other ungulates is uncertain. For example, neither pronghorn nor white-tailed deer display rump patches during copulation (Hirth and McCullough 1977).

Smythe (1970, 1977) proposed that rump-patch displays evolved as a pursuit-invitation signal to predators. Hence, a prey animal that had detected a predator could display its rump patch and thereby elicit the premature pursuit of a predator while the predator was still at a safe distance, which would lower the likelihood of a successful attack. A related hypothesis of pursuit deterrence was formulated by Woodland et al. (1980). This hypothesis holds that rump patches and associated behaviors advertise the ability of the prey to elude capture, and thereby discourage the predator from pursuing prey. Both pursuit-invitation and pursuit-deterrence hypotheses may be applicable to species like white-tailed deer, which have the ability to turn their rump patch "off and on" by raising or lowering their tail. Nevertheless, neither of those hypotheses offer an adequate explanation for large and permanent rump patches, such as those of bighorn sheep or North American elk, that cannot be "turned off." Similarly, Pipia et al. (2009) proposed that ungulates may signal to the predator that it has been spotted to eliminate advantages of a surprise attack. One potential shortcoming with these hypotheses is that it would be maladaptive for ungulates with permanent rump patches that cannot be concealed to continually elicit pursuit, indicate their presence to predators by signaling them, or to advertise their ability to escape—there might be circumstances where predators were encountered at close distances where signaling a predator would be disadvantageous (Coblentz 1980). Moreover, unless rump-patch displays are so costly that sick or otherwise infirm individuals cannot engage in the behavior, then "cheating" by disadvantaged prey would be expected, and predators would have to pursue prey continually to "test" their physical condition (Caro 1986). Antipredator benefits related to group cohesion in ungulates for tail-flagging by alarmed white-tailed deer, and the subsequent grouping of individuals in more open habitat, offers a viable hypothesis for alarm signaling in ungulates (Hirth and McCullough 1977). In those examples, habitat and predation risk combine to affect gregariousness.

Ungulates often employ a sequence of alert, alarm, and flight behaviors that reflect the proximity of predators and the attempts of ungulates to elude or evade those carnivores, as I documented in my research in the Cuyamaca Mountains of southern California (Bowyer et al. 2001a). For example, mule deer exhibit a series of behaviors that initially start with an alert posture with the head held high and ears directed forward toward a predator (or disturbance). That posture progresses into a stilted and exaggerated, stiff-legged walk, with ears held backward and no longer in an alert posture, as deer begin to move away from the predator. The stiff-legged gait is likely homologous with the alarm walk described for black-tailed deer by Stankowitch and Coss (2008). The stiff-legged gait escalates to a trot as deer begin moving farther away and they become increasingly disturbed. Fully alarmed mule deer begin stotting—using a bouncing gait with all feet landing near simultaneously that often involves abrupt changes in direction as deer flee, which may help them avoid obstacles or evade a pursuing predator (Caro 1986). Finally, if deer are approach too closely or are startled at close range, they may break into a gallop, which is the most rapid of their gaits (Bowyer et al. 2001a; Table 2). These behaviors often are contagious, with other group members also engaging in similar alarm postures and gaits. This sequence of alarm behaviors was identical for both approaches by coyotes and by a human observer, but deer reacted more strongly to the presence of coyotes (Table 2). Legal hunting of deer in this area was not permitted, but those animals clearly treated humans as if they were predators. Ungulates other than mule deer also engage in stotting behavior, which is thought to provide antipredator benefits (Caro 1986). White-

Table 2. Alert-alarm behaviors and gaits used by mule deer in response to approaches by coyotes and by a human observer in the Cuyamaca Mountains of southern California. Responses are ordered from least (Alert) to greatest (Gallop) alarm (from Bowyer 2001a).

| | Distance of disturbance from deer (m) | | | | | | | |
| | Coyote | | | | Human | | | |
Response of deer[a]	n	$\bar{x}$	SD	Range	n	$\bar{x}$	SD	Range
Alert	8	154	106	40–300	45	130	79	40–350
Stiff-legged walk	3	83	61	30–150	40	64	16	30–150
Trot	5	61	51	20–150	40	56	22	30–150
Stot	6	34	7	25–40	26	41	14	25–70
Gallop	3	28	13	15–50	4	14	6	10–20

[a] Responses of deer were scored only once for each encounter and included the closest approach and most severe response.

tailed deer, however, do not stot, but may escalate their alarm behavior and gallop away while tail-flagging (Hirth and McCullough 1977).

Coyotes in the Cuyamaca Mountains occurred in groups of 1–5 animals, with a mean of 1.33 coyotes/group for 216 groups. There was a tendency for coyotes not to approach or pursue deer when the number of deer per coyote was large, but to become more aggressive toward deer, sometimes attacking them, when the number of deer per coyote was reduced (Bowyer 1987). A larger group size of mule deer was influential in deterring attacks by coyotes.

Flight behaviors of ungulate groups are thought to confuse predators—the juxtaposition of fleeing pronghorn with piloerected rump patches may make the selection of an individual animal to pursue difficult (Kitchen 1974). Kruuk (1972) and Schaller (1972) observed that large carnivores that switched their pursuit from one ungulate to another had low rates of success. Another obvious antipredator strategy of some ungulates is to out run pursuing carnivores. The rapid acceleration and top speed of >45 miles/h (72 km/h) of fleeing pronghorn, with an ability to maintain those speeds over long distances, makes them an elusive prey (Kitchen 1974, O'Gara 2004b). Byers (1997) hypothesized that the swiftness of modern-day pronghorn resulted from them coevolving with 2 species of cheetahs (Acinonyx) during the North American Pleistocene—felids that undoubtably were capable of the rapid pursuit of prey. No extant large carnivore in North American can maintain such speeds, and Byers (1997) viewed this as an example of past adaptations being maintained over evolutionary time (i.e., ghosts of predators past). I offer an alternative suggestion that the swiftness of pronghorn still holds benefits for eluding predators. Coursing predators, including wolves, may pursue and harass prey in a series of chases lasting over long periods. I observed a wolf chase a caribou in a series of pursuits that lasted most of a day in interior Alaska. The caribou fled from the wolf in each encounter that I observed (I did not personally see them all, but my colleague

Victor Van Ballenberghe observed many interactions), but the caribou was never able to completely evade the wolf, and ultimately it was killed. I hypothesize that the speed and distance covered by pronghorn functions to break contact with modern-day predators and prevent extended pursuits by large coursing carnivores, which reinforces natural selection for speed and endurance. Moreover, Schaller (1968) observed that hunting cheetahs (Acinonyx jubatus) maintained top speeds for 219–328 yards (200–300 m) in pursuit of prey and, when they stopped, appeared exhausted. The hunting behavior of modern cheetahs (and presumably their ancestors) does not explain the ability of pronghorn to maintain high speeds over long distances to elude a predator that specializes in short pursuits.

An additional benefit of grouping for open land species may be an active defense against predators, such as the defensive stance of muskoxen (Gray 1987). Muskoxen increased their group size relative to the density of wolves (Heard 1992). Sinclair (1977) and Prins (1996) documented aggression by African buffalo (Synoorus oaffor) toward predators during attacks. Even so, less gregarious ungulates that stand their ground against predators also may lower their risk of being killed (Mech 1970, Bowyer 1987, White et al. 2001, Berger 2018). Ungulates also may harass predators as a defensive strategy (Berger 1978, Bleich 1999, Grovenburg et al. 2009).

Ungulates are divided into 1 of 2 categories related to the type of mother-young relationship that occurs during the postpartum period: *hiders* and *followers* (Lent 1974, Walther 1984). The hiding of young is ostensibly an antipredator strategy in which crypsis is used to conceal the young from predators during the first few days of life, a pattern typical of species inhabiting forests and woodlands. Neonates that move with their mothers soon after birth are termed followers, a pattern characteristic of open-land ungulates. Followers are thought to garner antipredator benefits of gregariousness, or sometimes from a group defense. Nonetheless, Green and Rothstein (1993) argued that this dichotomy was an oversimplification of mother-young spatial relationships. For instance, American bison (a gregarious open-land species) studied by Green and Rothstein (1993) in the Black Hills of South Dakota hid their neonates. Black-tailed deer (supposedly a quintessential hider) inhabiting the Trinity Alps of northern California followed their mothers and joined groups of deer 2 to 3 days following parturition (Bowyer et al. 1998a). Neonatal deer experienced an alarm bradycardia when disturbed, causing them to remain motionless or only flee short distances (Jacobsen 1979). This modality of alarm responses indicates that under particular circumstances gregariousness likely provides an additional antipredator behavior to remaining motionless and employing crypsis. I suggest that a combination of

habitat, types of predators, and the risk the predators pose plays a central role in determining the type of antipredator behavior exhibited in mother-young relationships. I question if the hider-follower concept should be linked directly to whether ungulates live in forests or open lands, and future research into that question should focus on variation in the spatial relationships between mother and young under risk of predation, recognizing that there may be variation in behavior even within species.

Ungulates may vary their group size, vigilance, foraging behavior, and habitat use in response to the hunting style (e.g., ambush or stalking versus coursing) of predators (Bowyer et al. 2001a, Atwood et al. 2009), which reveals a complex association of behaviors related to social groupings under preda-

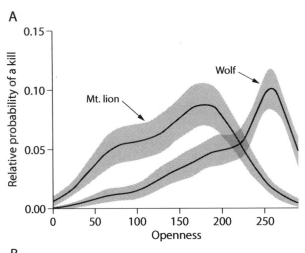

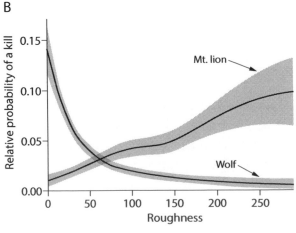

Fig. 15. Spatial distribution of elk (adult females and young) killed by wolves and mountain lions during winter in northern Yellowstone National Park relative to vegetation openness (*A*) and topographic roughness (*B*). Lines are fitted values with 95% confidence intervals (*shaded areas*). (Modified from Kohl et al. 2019)

tion risk. Eurasian lynx (*Lynx lynx*), which are stalking predators, killed smaller female red deer disproportionally to their availability compared with larger male red deer, but they did not differentiate between sexes of the smaller and less sexually dimorphic roe deer (Heurich et al. 2016). Pierce et al. (2000b) noted that mountain lions, which are also stalking predators, preyed more intensively upon young mule deer than did coyotes, which are coursing predators. Moreover, Kohl et al. (2019) demonstrated that wolves hunted primarily during the day, whereas mountain lions hunted mostly at night. Further, those authors demonstrated differences in kill sites for North American elk based on habitat openness and terrain ruggedness, further differentiating hunting styles of those large carnivores (Fig. 15). Male wildebeest (*Connochaetes taurinus*) killed by coursing predators exhibited more tooth wear than males killed by stalking predators; females demonstrated the opposite pattern (Christianson et al. 2018). Such patterns of tooth wear ostensibly reflect changes in foraging behavior in different areas resulting from sexual dissimilarities in susceptibility to various hunting styles of predators (Christianson et al. 2018). Clearly, predation is a critical aspect of the lives of ungulates, which holds important implications for sociality, risk of predation, and by extension, understanding sexual segregation (Fig. 4).

SUMMARY

This chapter sets forth the definitions for sexual segregation and deals with difference in morphology, physiology, and behavior related to this phenomenon. The chapter also considers the role of resources and predation risk in influencing segregation of the sexes.

1. The traditional definition for sexual segregation is the differential use of space or other resources by the sexes outside the mating season. Alternative views of sexual segregation involve differences in activities of the sexes of sexually dimorphic species, along with several other hypotheses related to social factors. Substantial confusion exists over specifically what constitutes sexual segregation, and the lack of a generally accepted operational definition is a primary hindrance in describing and understanding this phenomenon.

2. Ungulates that sexually segregate are sexually dimorphic (males larger than females). Two potential routes for the evolution of sexual dimorphism are competition within sexes for mates (sexual selection) and competition between the sexes for resources (natural selection). Increases in body size of males are related to increases in the degree of sexual dimorphism (Rensch's Rule). For ungulates, sexual selection offers the best explanation for sexual dimorphism, which is reflected in the disproportionate size of males and the horn-like structures they use in combat with competitors for mating opportunities.

3. Because of sexual size dimorphisms, the architecture of the mouth (especially incisor breadth), and its relationship to dietary differences in ungulates, was thought to cause differences in habitat use and thereby cause sexual segregation. Incisor breadth, however, differs little between sexes and modifications in the morphology of the mouth are likely a result rather than a cause of dietary differences.

4. Differences in the ratio of body size to the size of the digestive organ (rumen and reticulum) offer an explanation for niche partitioning among an array of ungulate species (the Bell-Jarman Principle). That relationship probably does not apply to body size differences between the sexes, however, and fails to provide an adequate explanation for sexual segregation and aggregation.

5. Differences in the digestive morphology and physiology between larger males and smaller lactating females can result in dietary

differences for ruminants. Males may obtain adequate nutrition on lower-quality, more abundant foods than those high-quality foods required by pregnant females, which can lead to sexual segregation (the gastrocentric hypothesis).

6. Distribution and quality of resources can modify sociality in ungulates, including the patchiness of habitats and how food items are dispersed and foraged upon.

7. A suite of life-history characteristics, morphology, and behaviors have been associated with ungulates living in open as opposed to closed habitats, including marked differences in group size. Group size also is constrained by the brokenness and steepness of terrain, especially for mountain-dwelling ungulates.

8. Predation risk plays a major role in gregariousness in ungulates. The probability of detection by predators can be reduced by antipredator behaviors in ungulates, including use of concealment or decreased probability of capture via vigilance, flight, use of escape terrain, or group formation. Females and young usually are more vulnerable to predation than males.

9. Risk of predation is thought to influence sociality of ungulates through the benefits of increased group size in open-land species; these benefits result from more eyes, ears, and noses with which to detect predators at distances where successful pursuits are less

likely. Individuals in large groups also reduce their probability of being selected as prey (dilution effects) as they become members of larger selfish herds.

10. Benefits also occur for forest-dwelling species from living in small groups. Noise and odors from large groups of ungulates moving through dense vegetation might interfere with detection of ambush or stalking carnivores, whereas small groups can be more secretive.

11. Some ungulates attempt to space away from predators, especially using sites for parturition that are at elevations above the typical distribution of predators or that occur in precipitous terrain that predators have difficulty negotiating.

12. Ungulates use specific alarm signals and gaits when confronted with predators, which may help elude or confuse predators, or may promote group cohesion.

13. Some ungulates may escape predation by swiftly fleeing or by offering an active defense against predators, both of which can be effective behaviors under particular circumstances.

14. Ungulates may adjust their behavior, feeding activities, and habitat use in response to the hunting style of predators (ambush or stalking versus coursing).

15. Understanding how factors affecting sociality differ between the sexes is a prerequisite for understanding sexual segregation.

3 | Temporospatial Patterns

W hy do the sexes of ungulates alter patterns of sexual aggregation and sexual segregation? The spatial distributions of the sexes on the landscape and how that arrangement changes over time are key components in understanding sexual segregation. Of particular importance is recognizing factors associated with the aggregation of the sexes during rut, and segregation around the time of parturition; those patterns are variable, however, including extending the period of sexual segregation into winter. Patterns of aggregation and segregation often are related to the types and sizes of social groups, and to the differing morphology of the sexes that reflects their divergent reproductive roles. Sexual segregation also is affected by the distribution of preferred habitats and forages, and by the differential vulnerability of sexes to predators. All of those factors are intertwined with the distribution of the sexes on the landscape. Indeed, many changes in patterns of behavior and group membership clearly are related to sexual differences in acquisition of resources and lessening risk of predation (Chapter 2).

Defining Social Groups

Although this book deals principally with concepts, few will be useful unless technical definitions and de-

tailed methods are operational to interpret the ideas. For the most part, I have minimized complex mathematical formulas but provided key references for those interested in a fuller treatment of the topics discussed. There are, however, technical details that are necessary to understand group living and sexual segregation.

I discussed the advantages and detriments of living in groups in Chapter 2, but thus far have not defined a group, which is not as simple or straightforward as might be expected (Kasozi and Montgomery 2020). Sociality means group living (Alexander 1974), and evaluating the complete range of sociality for groups of ungulates is essential to understanding their behavioral ecology. This approach requires discarding the dictionary definition of a group as 2 or more individuals. Solitary individuals are a critical element in the social systems of many ungulates, and lone individuals need to be considered along with larger groups to fully interpret their social behavior and organization (Monteith et al. 2007). Several studies of group size in ungulates have omitted this crucial element in assessing degree of sociality, including some studies that used analytical methods to test for sexual segregation (Chapter 4). In addition, a number of considerations exist in defining what constitutes a social group beyond just the number of animals contained therein (Fig. 16).

Attributes of Group Size

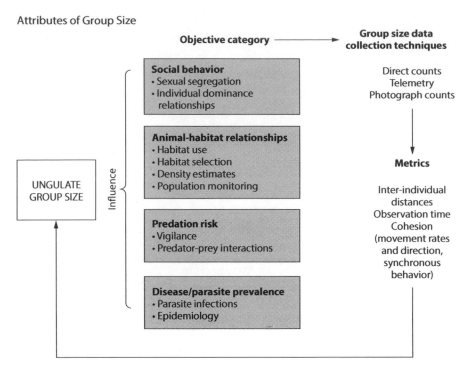

Fig. 16. A conceptual framework for estimating ungulate group size. Studies describing ungulate group sizes organized into 4 objective categories. The study objectives determine the field techniques for estimating size of groups. Field techniques used for group-size estimation determine the metrics employed and ultimately the group sizes reported by studies. Arrows depict the sequence of events that lead to observed group sizes of ungulates. (From Kasozi and Montgomery 2020)

Attributes of Groups

To include the compete spectrum of sociality from solitary to gregarious, I followed Siegfried (1979) and have used a distance criterion in defining a group as one or more individuals that were ≤50 m from their nearest neighbor, and that were apparently aware of one another, as evidenced by staring at other group members (Bowyer et al. 2001a, Bowyer et al. 2007). If questions arose as to whether an animal was part of a particular group, that individual was watched until it clearly joined the group or moved away. In practice, there was little difficulty in distinguishing groups using those criteria because groups tended to travel as relatively cohesive units (Bowyer et al. 2001a). Use of a particular distance (e.g., ≤50 m) between group members in defining a group is somewhat arbitrary but should bear some relationship to

the overall size of social groups. Longer distances may be appropriate for more gregarious ungulates (Kasozi and Montgomery 2020). Krause and Ruxton (2002) termed this model for understanding spacing of individuals within and among groups the "*elective group size concept.*"

Some groups of ungulates, such as those of Roosevelt elk (*C. e. roosevelti*), maintained relative constant group membership over time, especially herds of females and young, whereas others, including mule deer, exhibited less stable groups (Bowyer 1981, Bowyer et al. 2001a). Mule deer moved to and from woodlands into meadows where they foraged at dawn and dusk—groups were joined by additional deer as they ventured farther from cover into open areas (Bowyer et al. 2001a). Hence, I needed to determine whether nearby deer were part of the same group. Because group sizes of mule deer were not stable, I evaluated several attributes of groups to determine whether criteria I used provided a reasonable definition of a group. The mean (±SD) nearest-neighbor distance between 106 groups of mule deer encountered during the same observation period was 367 ± 280 m, whereas the mean distance across each of those groups was only 38 ± 32 m (with solitary deer excluded for obvious reasons). A clear pattern of grouping on the landscape was evident. In addition, there was greater synchrony of foraging within groups (75%) than among groups (58%), reflecting the cohesiveness and coordination of activities within groups. These assessments of distances and activities helped determine that my definition of a group was reasonable; this approach should be useful for studies when group size is variable yet necessary to investigate the ecology or behavior of ungulates.

Jarman (1974: appendix 2) proposed that "typical group size" be calculated to deal with high variance and skew in the distribution of groups. Typical group size was equal to,

$$\sum n_i^2 / \sum n_i,$$

where n_i is the number of animals in the ith group. He argued that this measure provided a more relevant metric that reflected the size of the group in which an animal found itself, as opposed to the mean size of all groups. Mean group size, however, offers a more standard metric that still provides interpretable data. In addition, means are widely used in the literature on group size, and typical group size would make comparisons among studies difficult. An alternative approach might be to provide the median and the interquartile distance as a measure of central tendency and variance for groups if their distribution was kurtotic or skewed, but again the opportunity for a comparative approach would be limited. Nonetheless, Weckerly (2020) used typical group size effectively in testing hypotheses concerning different-sized groups of male Roosevelt elk. Under most circumstances, I do not perceive a huge value in typical group size

over arithmetic mean group size, but I strongly suggest that, whatever measures are used, variance in size of groups or a graphic depicting number of animals in relation to groups of various sizes be provided.

Noteworthy patterns do exist between the number of social groups and the number of animals that compose those groups, allowing an evaluation of how individuals are distributed into those social units. Sinclair (1977) reported an inverse and curvilinear relationship (a negative binomial distribution) between group size and the number of herds for African buffalo. In addition, he noted that the relationship between group size and number of individuals was parabolic—most buffalo were in herds of intermediate size with comparatively few individuals in either exceptionally large or small herds. This distribution provides a clear picture of group living for individuals (Sinclair 1977).

Such patterns, expressed as percent of group sizes for ungulates, also may be manifest seasonally. Solitary individuals composed a large proportion of groups for mule deer but a comparatively small component of mule deer in groups of one (Fig. 17). Bowyer et al. (2001a) reported that seasonal changes in groups and the individuals they contain exhibited configurations that reflect differences in the concealment cover, quality of resources, and life-history patterns of mule deer. For example, concealment cover reached seasonal lows during winter, but growth of forage plants in meadows, resulting from autumnal rainfall, had started to occur in that montane ecosystem—both factors that can promote increases in group size (Bowyer et al. 2001a). Moreover, those mule deer rut in November and December, which promoted larger groups (Bowyer 1986a). Conversely, vegetative cover was high in spring and early summer, with parturition occurring in early June and July, circumstances that favored smaller groups of deer (Bowyer 1991). Some interannual variation also occurred, but those general patterns of grouping in deer were maintained (Fig. 17).

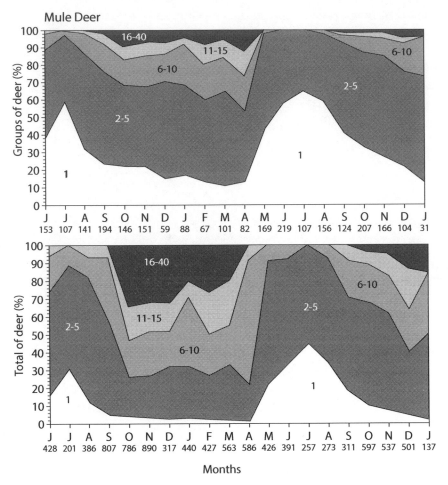

Fig. 17. Monthly changes in the size of 2,639 groups of mule deer (*above*) and 9,260 total deer composing those groups (*below*) in the Cuyamaca Mountain of southern California. Numbers within the graphs indicate categories of group size. Monthly sample sizes, which resulted from repeated observation of about 250 deer, are provided below letters corresponding with months. (From Bowyer et al. 2001a; originally published in *Alces*)

Types of Social Groups

Definitions for group membership can influence subsequent descriptions and analyses of group associations related to temporal changes in reproductive physiology and behavior of ungulates. Such changes hold importance for delineating and understanding periods of sexual segregation and aggregation (Bowyer 2004). Hirth (1977) developed a widely used method for categorizing social groups of ungulates. Social groups included mixed sex, adult male, adult female, yearling, or young. A mixed-sex group contained ≥1 adult female and ≥1 adult male but also could include yearlings and young. An adult male group included ≥1 adult male but also could contain other sex and age classes of deer except adult females.

Similarly, adult female groups contained ≥1 adult female and could contain other sex and age classes except adult males. Yearling groups were composed of only yearlings of either sex, and groups of young included only those animals (Hirth 1977).

Monteith et al. (2007) defined social groups of white-tailed deer with a modified procedure that involved "*solitary categorization*" to better reflect the large number of lone individuals. Social groups were categorized as female, solitary female, male, solitary male, and mixed sex. Female groups consisted of ≥2 adult or yearling females and no males but could contain young. Male groups contained ≥2 adult or yearling males with no females or young. Solitary female groups contained 1 adult or yearling female and could include young. Solitary male groups contained

1 adult or yearling male. Mixed-sex groups contained ≥1 adult or yearling male and ≥1 adult or yearling female and could contain young.

The categorization method of Hirth (1977) revealed a slight decrease in adult female groups and a slight increase in yearling groups of white-tailed deer over time (Fig. 18A). Adult female groups declined from 70% in early May to 50% in July. Yearling groups exhibited peaks of 40% in early May and mid-June. Groups of adult males declined from 10% to almost 0% from early May to parturition, and then increased to 10% by mid-June. No groups of young

were observed. Adult female and yearling groups composed 60.1% and 30.9% of observations, respectively. Despite obvious trends, no statistical change (P = 0.33) in group membership occurred with the categorization method of Hirth (Monteith et al. 2007; Fig. 18A).

The solitary categorization method for those same white-tailed deer indicated groups of solitary males increased from 0% in April to 15% in July (Fig. 18B). Likewise, male groups increased from 1% in early May to 5% of groups by June. Female groups declined from 40% in April to 10% in July, whereas groups of solitary

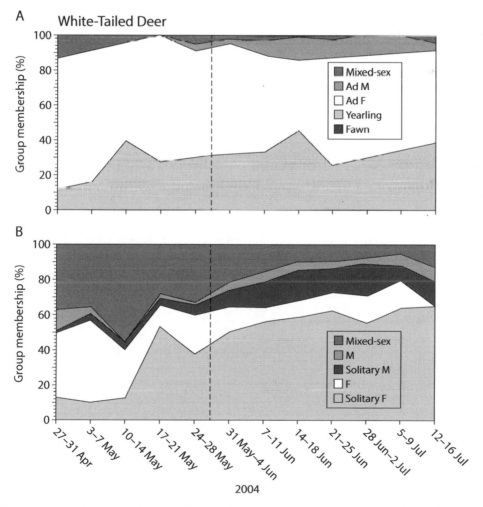

Fig. 18. Percentage of social groups of white-tailed deer based on observational sampling in Lincoln County, Minnesota. Dashed vertical lines represent peak parturition. (A) Group membership (e.g., group type) was classified with the categorization method of Hirth (1977). (B) Group membership was defined with a "solitary categorization" method. No fawn groups were identified as defined by Hirth (1977). (From Monteith et al. 2007)

females increased from 10% in April to 60% by July. In addition, female groups declined from 30% in early May to 10% in late May, with an associated increase in solitary females from 10% in early May to 50% in late May. Mixed-sex groups declined from 40% in early May to 20% in late May. Changes in group membership were dramatic and highly significant ($P < 0.001$) for the solitary categorization method (Fig. 18B).

Clearly, considering solitary females and males as a separate social unit in less social species such as white-tailed deer is appropriate and provides a more accurate representation of their reproductive behavior and degree of sociality than other methods (Monteith et al. 2007). For other less solitary and slower-developing ungulates, however, the method of Hirth (1977) remains useful, especially for species like bighorn sheep, where yearling males remain in female groups until body size approaches that of adult males (Bleich et al. 1997), and highly gregarious species, such as bison, where several age classes of males may occur in social groups, yet large males mate most often (Bowyer et al. 2007). The categorizations of groups can be meaningful where group types are used to identify periods of sexual aggregation (increases in mixed-sex groups and declines in single-sex groups) and segregation (declines in mixed-sex groups and increases in single-sex groups). Occasionally, however, sexual segregation can occur without obvious changes in the types of social groups, especially for less social species (Kie and Bowyer 1999), a point I will discuss later (Chapter 7).

Temporal Patterns of Sexual Segregation and Aggregation

Periods of sexual segregation and aggregation can vary across the year, as exemplified by bighorn sheep. Most mixed-sex groups occurred during rut, with the fewest of those groups occurring around the time of parturition, although desert-dwelling bighorn sheep have a protracted birthing period (Bleich et al. 1997; Fig. 19). Chiru (*Pantholops hodgsoni*) males and females migrate long distances to separate ranges during summer, reducing the occurrence of mixed-sex groups (Schaller 1998). This pattern of segregation and aggregation is typical for most sexually dimorphic ungulates, but nearly complete segregation or aggregation of the sexes is rare (Bowyer 2004). The uncommon occurrence of mixed-sex groups outside of rut may represent circumstances where the benefits of group living overshadow any detriments to the sexes from aggregating.

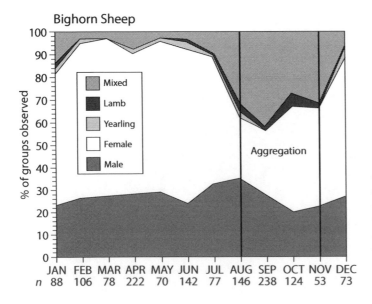

Fig. 19. Percent of social groups of desert-dwelling bighorn sheep illustrating periods of sexual segregation and aggregation. Note the large number of mixed-sex groups during aggregation, which corresponds with rut and the protracted period of segregation associated with the birthing and lower numbers of mixed-sex groups. (From Bleich et al. 1997)

For mule deer, a general pattern of sexual aggregation occurring during rut and sexual segregation being associated with parturition was evident. Those deer, however, also showed a second peak of mixed-sex groups in April, when there was increased growth of succulent plants in meadows on southerly exposures, and the sexes briefly concentrated in those high-quality areas (Bowyer 1984). White-tailed deer also may aggregate outside rut during harsh winter conditions related to effects of snow depth and the needs for thermal cover (McCullough et al. 1989, Beier and McCullough 1990). Black-tailed deer segregated around the time of parturition in montane meadows in the Trinity Alps of northern California, but some males and mixed-sex groups were present before males began migrating toward the Pacific Coast (Fig. 20).

For African buffalo living in more equatorial environments, a different pattern of segregation and aggregation occurs (Prins 1996). During an extended mating season of about 6 months, mature bulls alternate between mixed-sex groups (with females and young) and all-male groups (Turner et al. 2005). Males spend more time foraging, and regain body condition depleted from mating activities, while in those all-male groups (although at an increased risk

of predation) before returning to mixed-sex groups for mating (Hay et al. 2008).

The timing of sexual segregation and aggregation also can be variable within species and environments. Miquelle et al. (1992) studied sexual aggregation and segregation in Alaskan moose (A. a. gigas) in interior Alaska, with aggregation occurring most often during rut and segregation being most pronounced during winter. Nonetheless, those moose were also largely segregated during spring. Likewise, Bowyer et al. (2001b) reported that the sexes of moose inhabiting interior Alaska partitioned use of space during winter based on a habitat modification. Conversely, Oehlers et al. (2011) demonstrated that moose from coastal areas of southeast Alaska segregated principally during and following parturition in late May and June. Thus, the various environments inhabited by even the same subspecies of moose may lead to differing seasonal patterns of sexual segregation. Indeed, those examples of sexual segregation in moose all resulted from the differential use of habitat (or modifications thereof) or predation risk experienced by the sexes.

Studies that encompass an annual cycle in examining degree of sexual segregation (and aggregation)

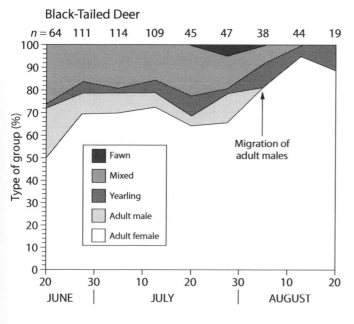

Fig. 20. Percent of social groups of black-tailed deer observed in the Trinity Alps of northern California during sexual segregation. Note that some mixed-sex groups are still present, although female groups predominate. (From Bowyer et al. 1996)

may provide important insights as to why the sexes segregate. Hypotheses that cannot account for such temporal differences in patterns of aggregation and segregation by the sexes on the landscape are not likely to offer an adequate explanation for sexual segregation in ungulates (Bowyer 2004). Hypotheses also should successfully address the adaptive significance of such changes in social groupings, an outcome that is not always clear-cut for some hypotheses for sexual segregation.

Spatial Patterns of Sexual Segregation

Patterns on the Landscape

A further consideration is that, in some species, such as bighorn sheep, males may move to and from ranges occupied by females to produce patterns of aggregation and segregation; adult males and females can occupy mountain ranges separated by ≥15 km when sexually segregated (Bleich et al. 1997, 2016). In other species, such as alpine chamois, females were segregated into high-elevation areas of high quality, whereas males occurred at low-altitude sites of poor quality—females joined males on low-elevation sites for rut (Unterthiner et al. 2012). In mule deer, females moved to ranges with water and more succulent forages during spring, thereby creating spatial patterns upon the landscape (Bowyer 1984). Hypotheses forwarded to explain sexual segregation also must deal with those contrasting patterns in the distribution of males and females on the landscape (too often patterns of movement by females are not fully considered). How, for instance, do differing activity patterns or social factors explain such variation in the manner and distances in which the sexes segregate from one another, especially for animals segregated into distant mountain ranges?

Some distinct patterns related to spatial separation of the sexes exist. Desert-dwelling bighorn sheep, for example, use mountain ranges with varying abundances of predators when segregated (Fig. 21). Whether measured by aerial surveys, time-lapse cameras, or walking transects in search of predator feces, adult males occupied mountain ranges with more large carnivores than did females (Bleich et al. 1997). During periods of segregation (which included parturition), females selected (use > availability) lower elevations and steeper and more rugged slopes with more open habitats than males. Females also occurred closer to water sources than males. Dietary overlap between the sexes also was affected by the distribution of bighorn sheep during sexual segregation (December–July). Moderate dietary overlap existed (61%–74%), with the lower value occurring in the second year of study when males consumed more shrubs and females ate more grasses (Bleich et al. 1997). Nonetheless, even small differences in diet quality can have a profound effect on productivity of ruminants (White 1983).

Where ground or aerial observation of groups of ungulates is difficult, telemetry locations also can be used to examine spatial patterns of sexual segregation and aggregation. Oehlers et al. (2011) employed multi-response permutation procedures (MRPPs) to analyze Euclidian distances between moose monitored simultaneously on the Yakutat Forelands in southeast Alaska. The average within-group distance, or delta value (the mean distance between all pairwise locations of each moose—male or female), was reported as a measure of spatial dispersion. Latitude and longitude were response (dependent) variables, and sex was the main effect (the grouping variable) used to test for spatial separation by month. MRPPs also were used to test whether a difference in the overall spatial distribution of males and females occurred with the excess function, which tests whether a particular group (male or female) could be obtained in a random draw from the joint distribution of the 2 groups. A clear and distinct pattern of delta values indicated moose were sexually separated starting in May and increasing in June (Fig. 22); late May and early June is the period of peak parturition for Alaskan moose (Bowyer et al. 1998b). Stewart et al. (2015) also used MRPPs to evaluate the timing of sexual segregation in North American elk.

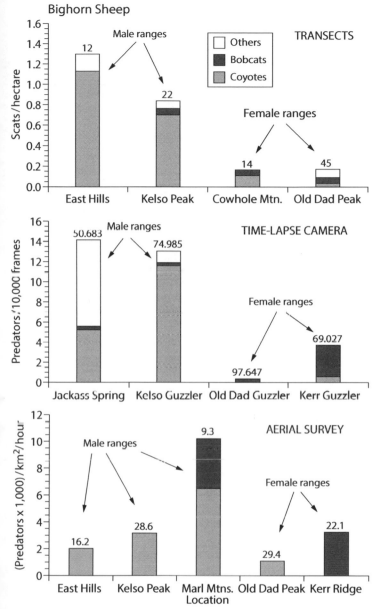

Bighorn Sheep

Fig. 21. Relative abundances of predators on ranges occupied primarily by mature male or female bighorn sheep in 5 mountain ranges in southern California, as determined by aerial surveys, time-lapse cameras, and feces collected along transects. Sample sizes are above bars and represent the number of transects, frames of time-lapsed camera film evaluated, and hours of helicopter time. Male and female ranges are indicated with arrows. (From Bleich et al. 1997)

Forage and Habitat Use, Selection, and Quality

Diets of the sexes may differ when they are spatially separated onto different ranges during sexual segregation. For example, Bleich et al. (1997) documented dietary differences between the sexes of bighorn sheep during sexual segregation. Such differences in composition and quality of diets between the sexes may be present but nevertheless may be subtle and sometimes difficult to detect. For instance, Bowyer (1984) noted that a primary difference in ranges occupied by male and female mule deer during segregation was the phenological stage of a preferred forage. Similarly, Beier (1987) reported slight, but significant, differences in foods eaten by the sexes of white-tailed deer during sexual segregation. Dietary differences between males and females also were

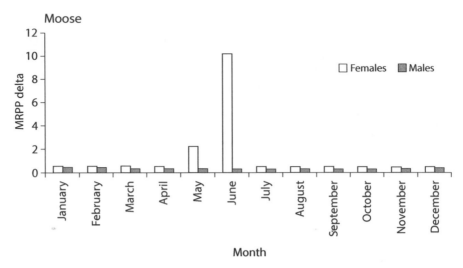

Fig. 22. Multi-response permutation procedure (MRPP; within group) values of delta by month for male and female moose on the Yakutat Forelands of Alaska. Delta values represent mean Euclidian distances between individual locations measured in decimal degrees, and they indicate strong sexual separation in June when females were giving birth. (From Oehlers et al. 2011)

reported for bison (Post et al. 2001, Berini and Badgley 2017); such dietary outcomes should be anticipated for most sexually dimorphic artiodactyls.

An important consideration in understanding dietary differences between the sexes, however, is whether the sexes prefer different forages or simply obtain different foods because the areas into which they separated have different availabilities of forage (Bowyer 2004). This outcome has implications for testing hypotheses concerning the role of body size, reproductive demands, and rumen-reticulum morphology and physiology in sexual segregation. Was the observed difference in diets of the sexes simply the result of differing availability of forages, or was the need for differing diets based on morphology and physiology of the sexes? A point that is often overlooked (Bowyer 2004).

A number of methods for evaluating resource selection exist (Johnson 1980, Boyce 2006, Thomas and Taylor 2006, Aho and Bowyer 2015a, and many others). Basically, all methods compare the use of a resource with its availability (Manly et al. 2002). The standard approach uses the selection ratio—the proportional use of a resource divided by its propor-

tional availability (use/availability). Within this framework, resources with a selection ratio equal to 1 are used in direct proportion to their availability, those with values >1 are "selected for," and those with values <1 are "avoided" or "selected against." Several new methods exist for calculating confidence intervals for selection ratios to determine if values differ significantly from 1 (Aho and Bowyer 2015a; http://cran.r-project.org/web/packages/asbio/index.html, accessed 20 April 2021), which I will not elaborate on herein but are worth examining if employing this approach. A critical issue is that sampling units used to quantify use and availability of habitats cannot overlap substantially. This circumstance may occur when multiple scales are used to investigate selection. As the size of sampling units are increased, overlap in used and available sites may grow to the point where it is not possible to discriminate habitats that are selected from those that are avoided (Bowyer and Kie 2006).

Importance values (proportional use × proportional availability) also add a useful element to studies of resource selection (Bowyer and Bleich 1984, Manly et al. 2002). The value of this metric is that it identifies commonly available resources that are nonetheless critical components in the ecology of an animal (Stewart et al. 2010). Such resources might not be identified as being "selected" in comparisons of use relative to availability (use/availability), especially if a critical resource was very common. Conversely, a rare resource might be highly selected with only limited use but be of overall low ecological value to animals because of its rarity. Importance values deal with those eventualities. A new method exists to calculate confidence intervals for importance values (Aho and Bowyer 2015b; http://cran.r-project.org/web /packages/asbio/index.html, accessed 20 April 2021). Use of that method likely will provide new insights into the value of resources for ungulates during periods of sexual segregation and aggregation.

Diet composition and quality are essential components for assessing sexual segregation with respect to different ranges occupied by the sexes. A plethora of methods exists to estimate composition and quality of ruminant diets (Church 1969, Robbins 1983, Hudson and White 1985, Van Soest 1994, Barboza et al. 2009, and others). I will address only a few of the common and noninvasive methods herein. Microscopic methods can be used to index the composition of diets from fecal samples by examining unique patterns of plant cells (Sparks and Malecheck 1968). This *microhistological technique* often includes correcting for the digestibility of various forages (Fitzgerald and Waddington 1979) with in vitro dry-matter digestibility (IVDMD) methods (Tilley and Terry 1963). Although providing a noninvasive method to estimate diets, microhistology still results in a considerable fraction of unidentified plant fragments in fecal samples, requiring the assumption that unidentified fragments occurred in the same proportions as those that were identified. Moreover, some plants cannot be identified to the taxonomic level of species.

Genetic techniques for determining diets of herbivores improve the accuracy of estimates obtained from fecal samples. DNA barcoding offers a modern approach that holds promise for obtaining detailed information on the diets of ungulates from their fe-

ces, with the ability to classify plants to species level (Valentini et al. 2009, Kowalczyk et al. 2019). Moreover, genetic methods exist for accurately identifying the sex of large herbivores from their feces (Brinkman and Hundertmark 2009), which is especially germane for studies of sexual segregation.

Stable isotope analysis, which can be applied to a variety of body constituents including fecal samples, has been employed to examine diets and trophic levels of mammals (Ben-David and Flaherty 2012 for review). Several elements occur in more than one stable form, termed isotopes. Carbon, for instance, occurs in a lighter form (^{12}C), which has an atomic mass of 12; the heavier form (^{13}C) possesses an atomic mass of 13. The behavior of isotopes in chemical reactions varies because of their different physical properties related to atomic mass, which leads to variation in the ratios of heavy to light isotopes in organic complexes. The ratios of heavy to light isotopes, for instance ^{13}C:^{12}C or ^{15}N:^{14}N, typically are expressed in relation to a standard and denoted as δ (del), which is the heavy form of the element (e.g., δ^{13}C or δ^{15}N). Isotopic signatures are the result of the ratios of heavy to light isotopes of the substrates organisms use, and of the physiological processes that occur in assimilating those substrates and eliminating their products (Ben-David and Flaherty 2012). Indeed, a multitude of potential climatic, dietary, and trophic interactions can influence stable isotope ratios (Ben-David and Flaherty 2012).

Stable isotope analysis has been employed to understand dietary niches for species of ungulates (Ben-David et al. 2001) and has provided new insights into both diets and trophic ecology of large herbivores (Stewart et al. 2003a). This methodology also was applied to differences between the sexes of American bison during periods of sexual aggregation and segregation (Berini and Badgley 2017). I believe that stable isotope analysis, when coupled with other information on diet and use of space, will become a powerful tool in investigations of sexual segregation. Indeed, isotopic niche size and degree of overlap for species or sexes can be quantitatively examined with a novel spatial metric (Eckrich et al. 2020; https://www.rdocumentation.org/packages/rKIN/versions/0.1, accessed 20 April 2021).

Fecal nitrogen has been used widely as a noninvasive index to diet quality for herbivores, including assessing differences between the sexes during periods of sexual segregation. This technique provides an index to variation in diet quality over time for many situations (Leslie et al. 2008). Nonetheless, recent experimental evidence indicates fecal nitrogen is not a reliable index to differences in diet quality between lactating females and adult males, and should not be used to assesses differences between the sexes of ruminants during parturition or for weeks thereafter

(Monteith et al. 2014a). Because of modifications of their digestive tract to accommodate the nutritional demands of lactation (see Fig. 8), lactating females are especially efficient at adsorbing nitrogen from forage, which is reflected in lower values of nitrogen in their feces than for nonlactating females or males (Fig. 23). Moreover, that pattern persists for both low- and high-quality diets until young are weaned (Fig. 23). Those feeding trials did not use diverse or natural forages, and outcomes from this research do not mean that females would prefer to use low-quality food if better forage was available. Nonetheless, this experimental study offers compelling evidence that fecal nitrogen will not suffice for de-

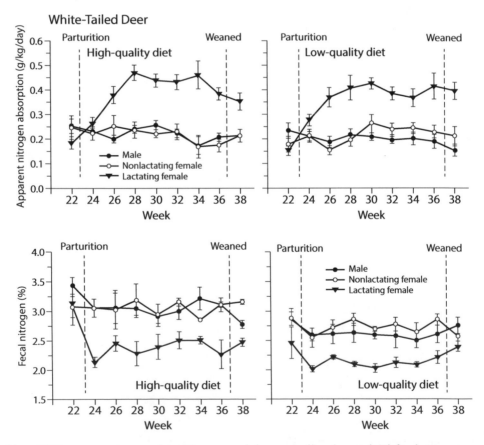

Fig. 23. Mean (±SE) apparent nitrogen absorption per week (*upper panels*) and mean (±SE) fecal nitrogen concentration per week (*lower panels*) of adult male, nonlactating female, and lactating female white-tailed deer in 2 treatment groups: high-quality diet and low-quality diet. Dashed lines represent the time of parturition and weaning of young for lactating females; South Dakota State University. (Modified from Monteith et al. 2014a)

termining differences in diet quality between lactating females and adult males during periods of sexual segregation. Moreover, this finding renders all comparative studies using fecal nitrogen to draw conclusions concerning sexual segregation in ruminants moot.

Several satellite-derived indices of habitat quality also are available, including the normalized difference vegetation index (NDVI), which reflects primary productivity and greenness of vegetation (Pettorelli et al. 2011). This metric allows comparisons of vegetative conditions on differing areas and can help judge the quality of habitat for herbivores over time (Lendrum et al. 2014). Note the increasing values of NDVI with declining snow depth and increasing temperature relative to the migration of mule deer from winter to summer range (Fig. 24). Monteith et al. (2011) provided similar information on migration of mule deer in the Sierra Nevada of California, but they also considered the physical condition and age of migrants in explaining those patterns. Care should be used in interpreting NDVI, however, because that metric may not reflect vegetation green up or quality when there is a substantial coniferous overstory (Chen et al. 2004). NDVI also may not be linearly related to changes in forage quality or quantity (Johnson et al. 2018). Moreover, all plants that are green are not necessarily nutritious food for ungulates. With those cave-

ats in mind, however, NDVI should be a useful tool in evaluating ranges of males and females during periods of aggregation and segregation, provided the sexes are not partitioning space at too fine a scale. The enhanced vegetation index (Kawamura et al. 2005) also has been used effectively to quantify vegetation quality for North American elk (Long et al. 2014).

Density Effects

Stewart et al. (2015) used MRPPs to examine spatial patterns of the sexes for North American elk in an experiment involving varying population density of those large herbivores on 2 large (842 and 260 ha) but adjacent and similar areas in northeastern Oregon. Elk were apportioned into the 2 areas, which were contained within an ungulate proof fence. Hunting of elk on those areas was not permitted. The low-density site contained 4 elk/km[2], whereas the high-density area was 20 elk/km[2] (Fig. 25). The duration of sexual segregation was 2 months longer in the high-density population (note the large delta values for females) and likely was influenced by individuals in poorer nutritional condition from intrasexual competition, which resulted in later conception and parturition, than for elk at low density. Males and females in the high-density population overlapped more in space but less in selection of resources

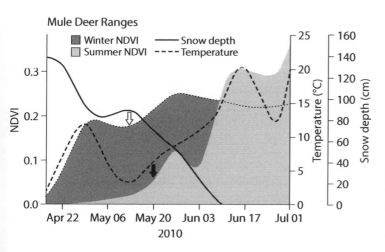

Fig. 24. Weekly change of the normalized difference vegetation index (NDVI) on winter and summer range of mule deer, mean weekly temperature, and mean weekly snow depth during spring in the Piceance Basin of Colorado. Arrows represent mean departure of deer (*white arrow*) from winter range and arrival date (*black arrow*) at summer range. The dotted line indicates winter NDVI. (From Lendrum et al. 2014)

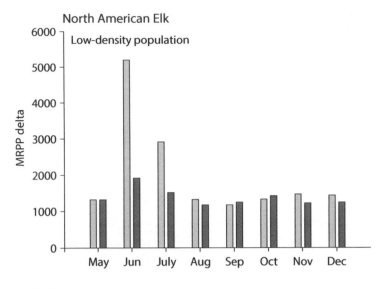

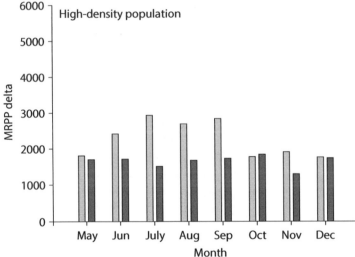

Fig. 25. Multi-response permutation procedure (MRPP) showing within groups values for delta by month for adult females (*light gray bars*) and adult males (*dark gray bars*) for an experimental manipulation of low-density (*top*) and high-density (*bottom*) populations of North American elk on the Starkey Experimental Forest and Range in northeastern Oregon. Delta values represent mean distance between individuals in each group measured in meters. (Modified from Stewart et al. 2015)

than in the low-density population, thereby exhibiting greater partitioning of resources at high versus low densities (Stewart et al. 2015). In another experimental study, McCullough (1979) noted that the sexes of white-tailed deer in southeastern Michigan increased their partitioning of resources at high compared with low densities (i.e., more overlap in space resulted in greater niche partitioning of habitats).

There are several important implications of these experimental approaches to understanding sexual segregation. All of the manipulations resulted in strong evidence of more resource partitioning at high density where sexual segregation was less pronounced. This outcome strongly indicates that niche dynamics of the sexes are involved in sexual segregation. Moreover, Kie and Bowyer (1999) documented that large increases in degree of sexual segregation occurred without marked changes in the types of social groups, which indicates that the manner in which social groups form and dissolve may not be strongly related to sexual segregation under some conditions. I will elaborate on these patterns in Chapters 6 and 7.

SUMMARY

This chapter defines the attributes and types of social groups. Both temporal and spatial patterns of segregation on the landscape are considered, including effects of forage and habitat use, selection, and quality. Effects of density are also discussed.

1. Defining social groups requires discarding the dictionary definition of a group as 2 or more individuals. Solitary individuals are a critical element in the social systems of many ungulates, and lone individuals need to be considered along with larger groups to fully interpret their social behavior and organization. Spacing between and within groups and their synchronous activities can be used to verify definitions of groups.

2. Seasonal changes in groups and the individuals they contain exhibited configurations that reflected changes in the concealment cover, quality of resources, and life-history patterns of many ungulates. Groups typically are categorized as mixed sex, adult male, adult female, yearling, and young. Treating solitary individuals as a separate category may provide further insights into changes in patterns of grouping, especially for less gregarious species.

3. Prominent seasonal patterns for many ungulates serve to illustrate the manner in which the sexes segregate and aggregate. Sexual segregation is seldom all-inclusive. Some sex and age classes behave differently from adults in their social groupings and in their spatial use of landscapes.

4. Sexual segregation is most pronounced during and following parturition, although sexual segregation may continue into winter in some species; sexual aggregation is most common during rut.

5. Variation exists in which sex moves to bring about sexual segregation. In some species, males move to female ranges for rut (i.e., aggregation); movements by males away from females at parturition produces segregation. For other species, just the opposite occurs. Depending on the species, sexual segregation can occur at large scales, including different mountain ranges, or at very fine scales of a few hectares. Such differences appear to relate to resource quality and risk of predation.

6. Differences may exist in the manner in which the sexes select food and habitat when segregated. Selection (use/availability) and importance (use × availability) of foods or habitats must be considered rather than just use, which may reflect differences only in the availability of forage in the areas into which the sexes segregate.

7. A number of methods exist to determine the composition of diets in ungulates. One noninvasive technique (microhistological analysis) examines microscopic configurations of plant fragments in feces. This method is being replaced by genetic techniques (DNA barcoding) that likely provide more accurate estimations of diet from fecal samples.

8. Stable isotope analysis, which can be applied to a variety of body constituents including feces, has been employed to examine diets and trophic levels of ungulates. This technique can be used to examine niche dynamics of the sexes during periods of segregation and aggregation.

9. Fecal nitrogen has been used widely as a noninvasive index to diet quality for ungulates, including assessing differences between the sexes during periods of sexual segregation. Experimental evidence indicates fecal nitrogen is not a reliable index to differences in diet quality between lactating females and adult males, and it should not be used to assess differences between the sexes of ruminants during segregation for weeks following parturition. Lactating females are especially efficient at absorbing nitrogen from forage, which is reflected in lower values of nitrogen

in their feces than for nonlactating females or males, even when initial diet quality was high.

10. Several satellite-derived indices of habitat quality are also available, including the normalized difference vegetation index (NDVI), which reflects primary productivity and greenness of vegetation. This index can be used to test for differences in the quality of ranges occupied by male and female ungulates during sexual segregation.

11. Experimental studies manipulating density of ungulates have provided additional insights into sexual segregation. Elk extended the period of sexual segregation for 2 months at high compared with low population density. Males and females in the high-density population overlapped more in space but less in selection of resources than in the low-density population, thereby exhibiting greater partitioning of resources at high versus low densities. A similar pattern occurred for a manipulation of population density in white-tailed deer.

12. These manipulations resulted in strong evidence of more resource partitioning at high density where sexual segregation was less pronounced. This outcome strongly indicates that niche dynamics of the sexes are involved in sexual segregation.

4 | Detection and Measurement

Detecting Sexual Segregation

How is sexual segregation detected and measured? Detecting sexual segregation can involve a plethora of variables and conditions, many of which are not independent (Table 3). Moreover, many of those environmental variables are complex and require the need for distance measurements (distance to roads, distance to concealment cover or escape terrain, etc.) and landscape metrics (fragmentation, patch size, edge density, etc.). Unfortunately, the suite of environmental variables chosen often depends on the definition of sexual segregation used. Because there is no standardized or generally agreed upon definition for sexual segregation (Barboza and Bowyer 2000), researchers use a quagmire of methods and variables to address sexual segregation across an array of environmental conditions (see Table 1). There have been few attempts to integrate those variables into hypotheses that lead to a coherent understanding of sexual segregation. A difficulty that arises is that many of the causes proposed for sexual segregation are not independent of other life-history characteristics of ungulates and may confound correlation with causation.

I believe there are several organizing principles for choosing variables useful to measuring and de-tecting sexual segregation in ungulates. The first is winnowing out some purported causes of why the sexes spatially segregate (I will elaborate on this in Chapter 5). Hypotheses that do not consider the adaptive significance of factors thought to promote separation of the sexes are unlikely to add much to understanding the evolution of sexual segregation. How benefits have accrued to the sexes as a result of behaviors leading to sexual segregation is critical to understanding this process. How those behaviors are maintained across a wide array of species via natural selection is the linchpin for recognizing which variables to measure. Mechanistic explanations without adaptive justifications are more likely to be correlates than causes of sexual segregation, and many methods used to detect sexual segregation are not designed to reveal its cause.

Another key element for sorting among hypotheses for sexual segregation (and consequently the variables to be considered) is whether variables are linked uniquely to degree of sexual size dimorphism. Among ungulates, it is primarily sexually dimorphic ruminants that sexually segregate (Chapter 2). If hypotheses focus on factors that also occur among other ungulates where sexual segregation is not known, then those factors may be related to variables causing sexual segregation, rather than promoting it

Table 3.   Variables potentially used in the study of sexual segregation

Related to the Animals
    Number and density
    Sex, age, size, and dominance status
    Degree of sexual dimorphism
    Mating system and degree of polygyny
    Number of groups
    Group size
    Group type (sex and age composition)
    Spacing among and between groups
    Temporal cohesion of groups
    Diel activities (e.g., standing, traveling, feeding, drinking, bedding, ruminating)
    Social behaviors (e.g., mating, maternal, aggressive-submissive, play)
    Alert and alarm behaviors

Related to the Environment
    Scale of measurement
    Weather and climate
    Habitat abundance and quality
    Habitat structure and configuration
    Forage abundance and quality
    Elevation, topography
    Rivers, streams, lakes, and ponds
    Density and species of large carnivores
    Hunting style of large carnivores
    Alternative prey for large carnivores
    Human developments and activities

themselves. Such factors are known as *"lurking variables."* A lurking variable is correlated with the variable of interest (e.g., degree of sexual segregation) but is extraneous and not its cause (Bowyer et al. 2020b). Lurking variables obscure the actual relation between sexual segregation and independent variables of interest and may cause misinterpretation of results and misleading conclusions.

Separating proximate from ultimate causes of sexual segregation is essential to understanding which variables play the primary role in explaining this behavior. Ultimate factors are ones that can explain sexual segregation across a wide variety of ungulates that engage in this behavior. Proximate causes are sometimes mistaken for ultimate causes of sexual segregation because they are seldom differentiated adequately from other explanations that do pertain across species. Proximate causes, and the variables that underpin them, may be red herrings diverting attention and

resources from efforts to understand the ultimate causes of sexual segregation. Consequently, we should focus our attention on explanations and the underlying variables that are supported by evidence from a wide range of species and conditions.

Variables proposed to explain sexual segregation should exhibit annual variation; they must have the capacity to explain why the sexes segregate and aggregate across seasons. Biggerstaff et al. (2017), for example, tested hypotheses concerning sexual segregation in white-tailed deer by sampling during a single month (August), which did not coincide with either peak segregation or aggregation. Moreover, sexual segregation cannot be inferred solely from differences in sex ratios of adults. Ungulates display skewed sex ratios because mortality of males typically exceeds that of females (Toïgo and Gaillard 2003). For instance, on a year-round basis, <20 adult males occurred per 100 adult females in an unhunted population of mule deer (Bowyer 1991). Changes in sex ratios in space and time are necessary to propose sexual segregation; a skewed sex ratio alone is not sufficient.

If social factors are invoked to explain sexual segregation, then those variables must exhibit seasonal variation and have capability to explain why the sexes use different spatial components of the landscape. Indeed, ecological explanations for sexual segregation must be able to integrate space, diet, and scale to understand this process.

Measuring Sexual Segregation

Segregation Coefficients

A systematic measure to detect and assess the strength of sexual segregation is essential for understanding this phenomenon. A number of approaches have been used to measure sexual segregation. Conradt (1998b) proposed to measure the degree of social segregation with a *segregation coefficient*:

$$SC_{social} = 1 - \frac{N}{X \cdot Y} \cdot \sum_{i=1}^{k} \frac{x_i \cdot y_i}{n_i - 1},$$

where x_i is the number of males in the ith group; y_i is the number of females in the ith group; n_i is the groups size of the ith group ($n_i = x_i + y_i$); k is the number of groups with ≥ 2 animals; X is the total number of males sampled (excluding solitary individuals); Y is the total number of females sampled (excluding solitary animals); and N is the total number of males and females sampled ($N = X + Y$). $SC_{social} = 0$ if no segregation occurs and is equal to 1 if there is complete segregation of sexes. The method is thought to be independent of sex ratio, density of the population, and group size. Conradt (1998b) discusses several caveats related to distribution and size of samples necessary to use this approach.

SC_{social} varies from 0 to 1, but incremental changes in group composition of males and females do not produce a constant degree of change in the index—the relationship is curvilinear, and the power of the index to discriminate between values likely declines as aggregation increases. Indeed, this index does not deal as effectively with aggregation as it does with segregation (Bonenfant et al. 2007).

Another shortcoming of the technique is that sexual segregation can occur under conditions in which the segregation coefficient cannot be calculated (Bonenfant et al. 2007). The primary weakness of the method, however, is that it does not consider solitary animals (Bowyer 2004), which are an important component in the social organization of some ungulates (Monteith et al. 2007; Chapter 3).

Conradt (1998b) proposed a measure of the degree of *habitat segregation*, which was similar in many respects to the measure of the degree of social segregation:

$$SC_{habitat} = 1 - \frac{M}{Z \cdot W} \cdot \sum_{i=1}^{l} \frac{z_i \cdot w_i}{m_i - 1},$$

where z_i is the number of males in the ith habitat type; w_i is the number of females in the ith habitat type; m_i is the number of males and females in the ith habitat type ($m_i = z_i + w_i$); l is the number of habitat types that are used by ≥ 2 animals; Z is the total number of males sampled; W is the total number of females sampled; and M is the total number of males and females sampled ($M = Z + W$). Solitary animals are included in this measure, but only if they overlap in habitat use with other animals. This measure of habitat segregation reflects habitat use and not selection, which poses difficulties in interpreting the causes of overlap by the sexes in various habitats. Habitat use is biased by differences in habitat availability, which can lead to differences in use of habitats even when the sexes select those habitats in an identical fashion (Chapter 3). Moreover, habitat selection of ungulates likely would change with population density. In addition, dietary differences between the sexes are not accounted for in this metric; recognition that the sexes may obtain different diets in the same habitat is not considered.

Conradt (1998b) also developed an equation to measure *spatial segregation*:

$$SC_{spatial} = 1 - \frac{C}{A \cdot B} \cdot \sum_{i=1}^{r} \frac{a_i \cdot b_i}{c_i - 1},$$

where a_i = the number of males in the ith grid; b_i = the number of females in the ith grid; c_i = the total number of animals in the ith grid ($c_i = a_i + b_i$); r = the number of grids with ≥ 2 animals; A = the total number of males sampled, excluding males that are not alone in a grid; B = the total number of females sampled, excluding females that are not alone in a grid; and C = total number of males and females sampled ($C = A + B$). This approach addresses spatial segregation by sampling on grid squares. No recommendations are offered, however, as to the size or placement (i.e., randomization method) of grids, and there is no consideration of patch sizes of habitat. Nonetheless, this method offers an inventive manner in which to simultaneously assess social, habitat, and spatial segregation. Weaknesses of the technique are that too little consideration is given to the role of solitary individuals in the social systems of ungulates, that it fails to incorporate an understanding that resource-based segregation can result from dietary differences within habitats, and that habitat use rather than selection is modeled; mathematical difficulties also exist in the formulation of these indices. Moreover, as applied, this method often employs a very simplistic definition of what constitutes habitat. Importantly, there is not a process offered to determine if differences between the sexes are statistically significant; there is no test against random associations of the sexes (Bonenfant et al. 2007).

As Bonenfant et al. (2007) noted, the degree of aggregation is not dealt with adequately, which hampers use of the modified formula proposed by Conradt (1999). Conradt (2005) suggested that if the value of the square root in her equation for the segregation coefficient became negative (and hence the coefficient undefined), that outcome was indicative of aggregation. Perhaps so, but this seems unnecessarily obtuse and provides no indication of the degree of aggregation

Conradt and Roper (2000) developed a model to predict the degree of social segregation in a population based on observed differences in activity synchrony in mixed-sex versus single-sex groups. Among the assumptions of this technique is that the less likely group members are to synchronize their activities, the more likely the group is to fragment, and different activity patterns between the sexes lead to such fragmentation. Again, this method fails to consider solitary animals. Moreover, the model measures a process thought to promote social segregation rather than the actual degree of segregation. Thus, differential activities between sexes are proposed to be a cost of joining mixed-sex groups. This technique does not consider which activities animals are engaged in, and there can be substantial differences, for instance, in animals that are feeding as opposed to standing, holding alert postures, traveling, or fleeing—all of which might affect group cohesion. Moreover, what is not considered is that other factors, such as predation risk or clumped distribution of resources, may be compelling reasons to maintain group cohesion and may override any cost of differences in activity between sexes (Bowyer et al.

2001a, Fortin et al. 2009). Indeed, sexual segregation can occur without changes in the types of social groups (Kie and Bowyer 1999).

Sexual Segregation and Aggregation Statistic

Bonenfant et al. (2007) developed a *sexual segregation and aggregation statistic* (SSAS) to address some of the potential shortcomings with the segregation coefficients of Conradt (1998b, 1999). Thus, by a slight modification of the segregation coefficient, SSAS was linked to a chi-square (χ^2) distribution:

$$\text{SSAS} = \frac{1}{N}\chi^2 - 1 - \frac{N}{XY}\sum_{i=1}^{k}\frac{X_iY_i}{N_i}.$$

N_i is the group size of the ith group ($N_i = X_i + Y_i$), X is the total number of males sampled, Y is the total number of females sampled, and N is the sum of males and females sampled. SSAS tests the null hypothesis of a random association between sexes against the alternatives of segregation or aggregation.

Because sample sizes may not be sufficiently large in studies of sexual segregation to meet assumptions of a χ^2 analysis, Bonenfant et al. (2007) recommend the use of permutation procedures to generate samples to test the significance of the null hypothesis or to construct confidence intervals; if multiple comparisons are performed, a Bonferroni correction to alpha (the critical value for the test of significance) should be used. Bonenfant et al. (2007) cautioned, however, that the assumption of independent animal movements among groups may not be met; consequently, they suggested using the group of associated animals instead of individuals as the sampling unit.

The use of SSAS, although an improvement over the sexual segregation coefficient, does not provide a direct measure of sexual segregation—this is a test of whether segregation or aggregation differs from a random association of males and females. This metric cannot be used as a measure of strength of segregation or aggregation (Bonenfant et al. 2007). Hence, concluding sexual segregation is stronger or weaker among months for a particular species, or between species of ungulates, from SSAS alone is not justified. As with other measures of sexual segregation and aggregation, SSAS is sensitive to effects of scale and should only be applied after such effects have been investigated. Finally, I do not believe SSAS should be presented without accompanying information on sexual segregation and aggregation, such as changes in the composition of mixed- and single-sex groups over time. To do so is equivalent to presenting inferential statistical tests without the descriptive statistics upon which they are based, which is an annoying practice.

Tests of Ratios and Proportions

If the purpose is to test for differences in sex ratios between areas or seasons, then several inferential tests exist. Such tests often are based on quadrat or transect sampling. A critical point that is often ignored is whether a particular test requires sampling with or without replacement. Bowden et al. (1984) developed a method for sampling sex ratios that could be adapted for dealing with sexual segregation. Their approach requires sampling without replacement (e.g., samples of the same individuals may not be repeated). Tests of proportions also are available to investigate sex ratios that allow sampling with replacement (Bowyer 1991). The assumptions of such tests allow individuals to be recounted any number of times; Bowden et al. (1984) cautioned, however, that the variance of such samples may be biased unless sample size was large. Also, the proportions must be independent (e.g., it is not valid to test for differences between sexes in a particular area or any other situation where the proportions sum to 1). Tests of proportions (Remington and Schork 1970:218), then, must be between areas or time periods (e.g., percentage segregated). This raises the question, however, as to what constitutes a segregated quadrat or the size (scale) of such samples. As I noted in Chapter 2, sexual segregation is seldom complete; some adult males and females co-occur during most seasons. If using quadrat-based sam-

pling, then some general (and often arbitrary) rule must be applied to all sampling quadrats. I previously have used ≥90% of one sex or the other (adult male or adult female) in each quadrat to categorize quadrats as segregated (Bowyer et al. 1996), which corresponds closely with declines in mixed-sex groups of black-tailed deer.

Multi-response Permutation Procedures

Another potential statistical method to evaluate the occurrence and strength of sexual segregation is a multi-response permutation procedure (MRPP) (Mielke and Berry 2001). MRPPs, including the use of excess groups (Mielke et al. 1983, Zimmerman et al. 1985), provide a powerful test for spatial differences between animals, which allows for comparisons not possible with other statistical procedures. MRPPs are distribution-free statistics that rely on permutations of data based on randomization theory (Mielke and Berry 2001)—there is no need to select a scale for sampling. These tests use Euclidean distances rather than squared deviations (variances) common to most other inferential statistics. Statistics based on simple Euclidean distances have greater power to detect shifts in central tendency among skewed distributions than do tests based on the more conventional use of squared distances. Indeed, MRPPs are especially sensitive to distributional changes, even when sample sizes are small (Nicholson et al. 1997). MRPPs also provide an appropriate graphical method of examining the strength of sexual segregation using delta values (see Figs. 22, 25). MRPP statistics are supported by the statistical package Blossom (https://rdrr.io/cran/Blossom/, accessed 24 January 2022).

Information-Theoretic and Machine-Learning Approaches

An information-theoretic approach selects the top model explaining ecological phenomena using Akaike's information criterion (AIC) (Burnham and Anderson 2002). AIC models do not result in an explicit test and potential rejection of hypotheses as do frequentist statistics. Further, there is difficulty in knowing if all important variables have been included in models—the top AIC model might not be best if important variables are omitted from analyses (Mac Nally et al. 2018). Arnold (2010) recommended cojoining AIC to frequentist confidence intervals to obtain tests of hypotheses (sensu Thalmann et al. 2015), which I believe provides the most useful scheme for employing those approaches.

Machine-learning algorithms, such as *random forests* (Breiman 2001), have advantages over other methods, especially with respect to dealing with large, complex data sets, and coping with interactions and missing data. The random forest algorithm is based on classification and regression tree analyses (De'ath and Fabricius 2000) and will generate reasonable predictions across a wide range of data, and it is especially useful for data mining. This technique has been extended for analysis of resource selection by ungulates (Heffelfinger et al. 2020). These are correlational analyses, however, and may contain potential biases from lurking variables if not adequately controlled (Chapter 3).

Effects of Scale

Scale is the most problematic concept in understanding sexual segregation. The failure to address scale when dealing with sexual segregation risks missing patterns on the landscape because the "wrong" scale (or perhaps no defined scale) was selected for measurement. Moreover, considerable variation exists in the scale at which sexual segregation occurs, and no one scale will suffice for all species or environments (Bowyer 2004). For example, the sexes of chiru segregate into ranges that are hundreds of kilometers away from one another (Schaller 1968). In bighorn sheep, the sexes may segregate into different mountain ranges that are >15 km apart (Bleich et al. 1997), whereas Bowyer (1984) noted that mule deer segregated into areas that were only separated by about 4 km. In other studies, sexual segregation in mule

deer was detected in quadrats of 64 ha (Main and Coblentz 1996), and for black-tailed deer in quadrats of 80 ha (Bowyer et al. 1996). White-tailed deer, however, segregated on an exceptionally fine scale of a few hectares (McCullough 1979, Kie and Bowyer 1999, Stewart et al. 2003b). Different ungulate species inhabiting a variety of environments exhibit differences in the scale at which sexual segregation occurs—a difficult circumstance to deal with when trying to detect sexual segregation. Moreover, I do not foresee how social factors could cause such variation in the spatial scale of sexual segregation without invoking ecological causes.

Habitat Selection, Home Ranges, and Scale

Ecological components related to sexual segregation include habitat selection, factors affecting the size of home ranges, and the scale at which ungulates respond to those variables. Habitat use is not a meaningful measure of sexual segregation because it is conditioned by habitat availability, which can differ for males and females during sexual segregation. Natural selection cannot operate on habitat use by the sexes; instead, it favors selection for or against various habitats via the reproductive success of individuals choosing particular habitats and avoiding others. Habitat selection can affect the size and arrangement of home ranges, which influences the size and composition of social groups, and thereby sexual segregation. In addition, habitat selection and home ranges are scale sensitive. Measures of habitat selection and the scale at which it occurs are necessary to understand sexual segregation. Below are a few examples to emphasize this point.

Strong relations existed between a suite of landscape metrics measured at different scales and home ranges of female mule deer in California (Kie et al. 2002). Landscape metrics such as edge density, mean shape index, and fractal dimension were inversely correlated with home-range size in deer across all spatial scales. Nonetheless, significant responses of other metrics such as patch size, patch size coefficient of variation, edge contrast index, and patch richness were scale dependent—evident at some spatial scales but not others. In addition, most of the variation in home-range size was explained by landscape metrics measured within a radius of 2,000 m around each location (Kie et al. 2002). Much of the variation in home-range size was explained by landscape scale and heterogeneity of habitats. Mule deer likely made decisions about habitat selection at scales larger than their home ranges. Home-range size will affect the number of conspecifics with which animals can associate, and ostensibly is related to factors influencing gregariousness and sexual segregation.

Other factors can affect the size of home ranges of male and female ungulates, including differences in body size, resource availability, and reproductive behavior (Viana et al. 2018). Male ungulates tend to have larger home ranges than females, but sexual differences in the size of home ranges per se are not evidence of sexual segregation. Males might contain numerous females within their larger home ranges, a pattern that need not result in differential use of space by the sexes.

In another example, the early winter distribution and density of female moose and habitat heterogeneity were examined in interior Alaska (Maier et al. 2005). Effects of vegetation type, topography, distance to rivers and towns, occurrence and timing of fire, and landscape metrics were considered in a spatial linear model used to analyze effects of independent variables organized at multiple scales. The analysis revealed that the densest populations of moose occurred closer to towns, at moderate elevations, near rivers, and in areas where fire occurred between 11 and 30 years ago. Relationships among environmental variables with moose density occurred at a scale of 34 km^2 but differed markedly among habitat characteristics; length of rivers, recent fires, elevation, and slope were selected at smaller scales from 15 km^2 to 23 km^2 (Maier et al. 2005). No single scale would have captured the relationships necessary to explain effects of habitat on moose

density. As noted previously, density of animals can play a role in the degree of sexual segregation.

In addition to home-range size and population density, effects of scale are also pervasive for studies of habitat section. Schaefer and Messier (1995) reported few changes in patterns of habitat selection in winter by muskoxen across changing scales. Muskoxen selected for a greater abundance of graminoids and less snow cover at successively smaller scales. Nevertheless, the overall consistency in habitat selection with changing scales of measurement was thought to mean that local levels of habitat selection occurred within one scaling domain (Schaefer and Messier 1995). Not all potential scales were measured, but this hierarchical approach was necessary to reveal the overall pattern. Moreover, a temporal component to understanding hierarchical habitat selection may be required. Schaefer et al. (2000) reported that at the scale of the year-round range, sedentary and migratory populations of caribou exhibited similar patterns of *philopatry* (the tendency to remain near or return to a particular area). At the scale of the seasonal range, however, sedentary caribou exhibited fidelity from calving to mating, whereas migratory caribou exhibited fidelity only during late autumn. Scale was critical to understanding patterns of ungulates on the landscape.

Rachlow and Bowyer (1998) examined a suite of biotic and abiotic variables to understand habitat selection by female Dall's sheep (*Ovis dalli*) during the birthing season in the Alaska Range of interior Alaska. Sampling was conducted at both the microscale (1 m^2) of the birth site and macroscales reflecting the overall landscape of lambing habitat (15–450 ha). Maternal females selected large areas with steep, rugged terrain coupled with microsites of level terrain for birthing and care of neonates. Failing to consider microsites would have produced a model that overlooked the need for small, level areas for parturition and rearing of young, and would have indicated a much greater extent of potential lambing habitat than what was suitable. Again, scale is a necessary component to address habitat selection and thereby to understand sexual segregation in ungulates, which tends to be most pronounced near the time of parturition (Bowyer 2004).

Birth-site selection also was studied for moose in interior Alaska (Bowyer et al. 1999b). Females selected (use > availability) sites for giving birth that were on southerly exposures with low soil moisture and high variability in overstory cover. Moose selected sites to give birth based on microsite characteristics rather than on broad types of habitat, which were used in proportion to their availability (Bowyer et al. 1999b). Cover of forage, especially willows (*Salix* spp.), was more than twice as abundant on birth than random sites. An inverse relationship between visibility and availability of forage indicated that female moose made tradeoffs between risk of predation and food in selecting sites to give birth. Simply examining habitats used for giving birth at the landscape scale would have missed the primary reasons for moose locating sites in particular areas (Bowyer et al. 1999b).

Oehlers et al. (2011) studied habitat selection by the sexes of moose in the Yakutat Forelands, in southeast Alaska, using multiples spatial scales (radii of 250, 500, and 1,000 m) around each moose location. The best model during all seasons for both sexes was at the 1,000-m scale. Habitat selection by the sexes differed during periods of segregation surrounding parturition and lactation, but it did not do so during rut when sexes were more likely to be aggregated (determined with MRPP; see Fig. 22).

Female moose selected areas with higher elevations, and likely steeper slopes, during spring and summer (when females were giving birth, nursing, and had young at heel) than males, ostensibly because of a lower risk of predation (Oehlers et al. 2011). Females selected for higher edge density than males during spring and summer, which supports a hypothesis of increased vulnerability of young to predators during this period (Bowyer et al. 1998b, Keech et al. 2000, Kunkel and Pletscher 2001), and the higher need for concealment cover by maternal females than for males (Molvar and Bowyer 1994). The landscape arrangement of forested and more open habitats (with considerably more forage) likely represented a tradeoff between risk of predation and the increased need for maternal females to acquire the resources to support the high costs of lactation. Likewise, females selected higher stream densities than males during spring and summer, perhaps to meet the requirements of lactating females for free water (Bowyer 1984), but also because riparian zones have high-quality forage to meet the nutritional requirements of lactating females (Stephenson et al. 2006). These results are consistent with several ecological hypotheses for sexual segregation but were evident only when considered at multiple scales.

Problems associated with scale in understanding sexual segregation are even more pernicious than in studies of habitat selection. Because of the nature of such studies, variation can occur with the number of males and females on a quadrat and with the size and placement of quadrats. Bowyer et al. (1996) assessed the degree of sexual segregation in a population of black-tailed deer inhabiting the Trinity Alps of northern California (Fig. 26). The size of individual sampling quadrats was increased from 16 to 192 ha by sequentially combining adjacent quadrats along a 9-km transect during the period when sexes of deer spatially segregated during and following parturition. Quadrates were categorized as segregated if they contained ≥90% males or females. The temporal scale was varied by examining quadrats weekly or combining quadrats over 5 weeks (thereby creating a larger sample size). Larger quadrats were more likely to contain adult male and adult female deer and, consequently, be categorized as aggregated (i.e., not segregated), whereas smaller quadrats tended to contain mostly one sex (≥90%) and therefore were categorized as segregated (Fig. 26). Thus, the longer sampling period resulted in less segregation than the shorter one (Fig. 26). Moreover, the pattern of sexual segregation was clinal, and nearly any degree of sexual segregation could be acquired by varying the size of sampling quadrats. Hence, sexual

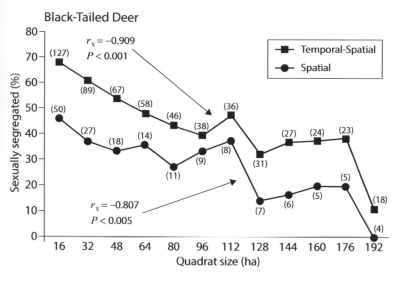

Fig. 26. Effects of quadrat size on measures of sexual segregation for black-tailed deer in the Trinity Alps of northern California. A quadrat contained >90% of either sex to be categorized as segregated. The temporal-spatial scale is analyzed by week, whereas the spatial-scale analysis combines 5 weeks. The number of quadrats sampled is in parentheses. (From Bowyer et al. 1996)

segregation was increased by either shortening the sampling interval or reducing quadrate size (Fig. 26). Bowyer et al. (1996) used a one-tailed runs test based on sequentially increasing size of adjacent quadrats to evaluate the largest quadrat that would still identify a pattern of sexual segregation, which was 80 ha.

Although this method helped select the most appropriate scale for assessing sexual segregation, the specific solution likely would be unique to the species and environment under study (Bowyer 2004). Many studies of sexual segregation fail to evaluate sampling scale or indicate at what scale samples were collected, making comparative studies of sexual segregation challenging and limiting the usefulness of some tests for why the sexes of large ruminants remain apart for a portion of the year (Bowyer 2004). Sampling at an inappropriate scale can prevent detection of sexual segregation, as noted by Kie and Bowyer (1999) for white-tailed deer.

A Synthetic Approach

One potential outcome is that other factors may have additive effects on sampling scale or may interact with scale to produce unexpected outcomes. Bowyer et al. (2002b) developed a landscape map from a study of mule deer conducted in the mountains of southern California (Nicholson et al. 1997). Four habitats were distinguished, and preferences of male and female deer were assigned to each habitat; the ecological carrying capacity (K) for that habitat was determined. The initial map, representing 1,000 ha, was termed coarse grained. A fine-grained map was developed by reducing pixel size in the coarse map to 25% of the original and combining the 4 copies of the reduced map into a new map of the area, with each quarter of the map representing an identical, reduced image of the original. This method resulted in 2 maps with considerable variation in landscape variables related to grain, including patch characteristics, shape indices, and measures of diversity and contagion (Bowyer et al. 2002b).

A high-density map (200 deer/1,000 ha) was generated by randomly assigning sexes of deer (35 males:100 females) to habitat patches under an ideal-free distribution until the K of a particular patch was reached. That same method was employed again to create a second map of low density (75 deer/1,000 ha). Sampling scale was varied by examining quadrats of 6 and 18 ha for sexual segregation. This design allowed evaluation of landscape grain, population density of deer, and sampling scale on measures of sexual segregation, which varied from 8.3% to 96.4% (Table 4). Logistic regression was used to determine which of those variables were influential in determining whether a particular quadrat was sexually segregated (≥90% of one sex).

Population density had the greatest effect on sexual segregation, with lower density resulting in greater sexual segregation, followed by sampling scale in which sexual segregation was greater at the smaller scale (Table 4). Landscape grain, however, did not significantly influence the model, perhaps because too few metrics of grain were considered, or because habitat selection by the sexes was not allowed to vary with patch size (Bowyer et al. 2002b). Moreover, no interactions occurred between grain, density, or scale. Those outcomes indicated, however, that consideration of how scale and population density of deer influenced sexual segregation should be important components of sampling designs, and they emphasized the point that other biological processes can either reduce or enhance scale-sensitive phenomena. Importantly, this model indicated that sexual segregation could occur without incorporation of social factors.

Mysterud (2000) subdivided ecological segregation into 3 broad categories: spatial, dietary, and habitat. His approach provides a framework for interpreting ecological segregation but has the drawback of separating interconnected processes (Bowyer 2004, Bowyer and Kie 2004, Stewart et al. 2011). Analyses based on those ecological subdivisions may erroneously conclude that sexual segregation does not occur when that assessment is based on a single category (e.g., space, diet, or habitat). Clearly, a more synthetic approach is needed. Such an approach is founded on a niche-based understanding of sexual segregation (Table 5) and proposes that the sexes of dimorphic ruminants behave as if they were separate but coexisting species (Kie and Bowyer 1999; Bowyer 2004; Bowyer and Kie 2004; Shannon et al. 2006a, b; Schroeder et al. 2010).

In an experiment involving white-tailed deer in South Texas, coyote numbers were reduced but not eliminated within a 391-ha enclosure. Densities of deer rose to high levels within the enclosure (77 deer/km^2), whereas outside the enclosure deer remained at moderate density (39 deer/km^2) (Kie and Bowyer 1999). No significant difference occurred in group sizes of deer between high and low density, and few density-related changes occurred in the types of social groups, especially those of mixed sex (Fig. 27); population density had little effect on deer group size or patterns of male and female associations. At moderate density, females and young made greater use than males of chaparral mixed-grass habitat with dense cover (where preferred herbaceous forage was less abundant), ostensibly a result of predator avoidance (Kie and

Table 4. Degree of sexual segregation in black-tailed deer resulting from sampling quadrats in 2 landscape grains, 2 population densities, and 2 scales of measurement. Twenty-four quadrats were sampled at the large scale (18 ha) and 72 quadrats were sample at the small scale (6 ha) (from Bowyer et al. 2002b).

Grain	Population density	Sample quadrat size (ha)	Quadrats with deer (n)	Sexually segregated (%)
Coarse	Low	18	13	84.6
		6	28	96.4
	High	18	24	8.3
		6	72	15.3
Fine	Low	18	14	64.3
		6	31	93.5
	High	18	24	12.5
		6	72	22.2

Table 5. A niche-based model for conceptualizing patterns of sexual segregation in polygynous ruminants (+ = overlap of sexes; − = separation of sexes). The model does not intend that overlap or separation be complete in any particular niche axis but indicates the general direction of outcomes. Other potential combinations of overlap and separation between sexes were judged to be inconstant with niche theory (e.g., overlap in space, diet, and habitat), or were otherwise infeasible (e.g., overlap in space but not habitat) (adapted from Bowyer 2004, Bowyer and Kie 2004). Herein we use a broad definition of habitat that can contain differences in resources.

Patterns of Sexual Segregation	Space	Diet	Habitat
I	+	−	+
II	−	+[a]	−[b]
III	−	−	+/−

[a] Some differences in diet would be expected because of allometric and life-history differences between sexes.

[b] Sexes potentially could use the same habitat at different locations.

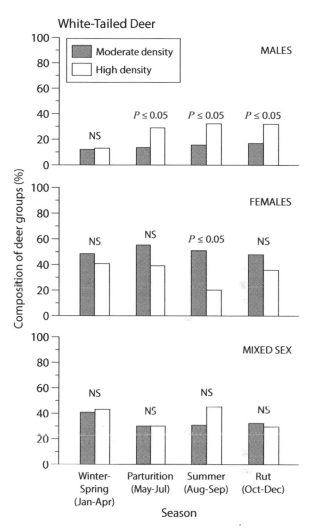

Fig. 27. Percentage of total groups categorized as male, female, and mixed sex for white-tailed deer at moderate (39 deer/km²) and high (77 deer/km²) densities resulting from an experimental manipulation of coyotes in South Texas. Significant P values for differences between moderate and high densities of deer are provided above each set of bars; NS = not significant. (From Kie and Bowyer 1999)

Bowyer 1999). Spatial segregation of the sexes was most evident at parturition at both moderate and high densities of white-tailed deer. Degree of sexual segregation, however, was significantly higher at moderate than at high density during all seasons (Fig. 28). Sexual segregation did not increase with increasing density of deer.

The principal component (PC) analysis in Figure 29 is designed to condense the large number of species that constitute deer diets into simpler and more understandable patterns, and to differentiate niche differences of deer between seasons, densities, and sexes (Kie and Bowyer 1999). Differences in diets between parturition and winter-spring seasons are demonstrated by 95% CI (confidence intervals) ellipses along PC1, with graminoids loading with large positive values and forbs with large negative scores, indicating a shift from cool-season grasses to a diet richer in forbs (Fig. 29A). PC2 was related to dietary differences between males and females at high density during winter-spring (Fig. 29A). At moderate density, diets of males were broad and encompassed the niche breadth of females entirely. At high density, diets of both males and females diverged markedly. Indeed, there was no overlap along PC2 between male and females (Fig. 29A). Grasses and browse loaded with large negative values along PC2, indicating that at high density during winter-spring, deer ate more grasses and browse and fewer forbs, a pattern most obvious among males at high population density (Fig. 29A). PC3 revealed

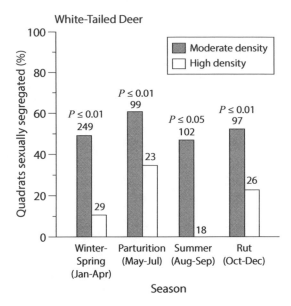

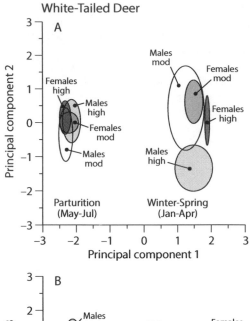

Fig. 28. Percentage of quadrats (2.9 ha) that were sexually segregated on a monthly basis at moderate and high densities of white-tailed deer in South Texas. Number of quadrats in each month containing deer were summed within seasons and are shown above each bar. P values are for differences between moderate and high densities of deer. (From Kie and Bowyer 1999)

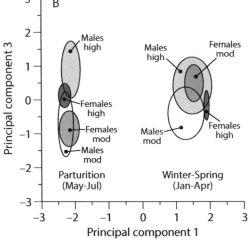

Fig. 29. Principal components 1 and 2 (A) and principal components 1 and 3 (B) for diets of white-tailed deer at moderate and high densities during winter-spring and parturition in South Texas. Ellipses are 95% confidence intervals. The first 3 principal components (PC) explained 66% of the variation in diets (PC1 = 40%, PC2 = 15%, and PC3 = 11%). (From Kie and Bowyer 1999)

dietary differences between males and females at high density during parturition (Fig. 29B). At moderate density, diets of males were varied and encompasses much of the diets of females, similar to diets in winter-spring as indicated by PC2. At high density, diets for both males and females changed markedly, and there was little overlap along PC3, indicating a greater reliance on those dietary items during parturition at high density (Fig. 29B). There is clear evidence that the sexes of white-tailed deer partitioned diets with respect to population density (which relates to space), with more overlap in diets at moderate density and less overlap at high density. Sexual segregation declined at higher density (more spatial overlap between sexes) but was compensated by an increase in diet partitioning, thereby avoiding competition between sexes.

Schroeder et al. (2010) studied sexual segregation in an endangered subspecies of bighorn sheep along the eastern slope of the Sierra Nevada of California

during winter. This research was conducted in the Mono Basin and in 3 areas making up the Southern Region (Mt. Langley, Mt. Baxter, and Wheeler Ridge). Bighorn sheep inhabiting the Mono Basin spent winters at high elevations, whereas those from the Southern Region occurred at low elevations during winter. Moreover, female bighorn sheep in both

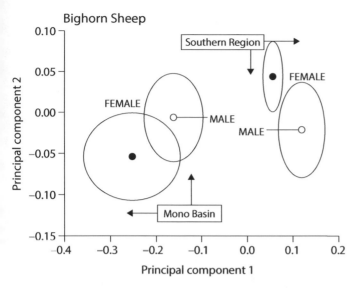

Fig. 30.   Principal components 1 and 2 for diets of male and female bighorn sheep in the Mono Basin and Southern Region of the Sierra Nevada of California during winter. Ellipses are mean principal components analysis scores and 95% confidence intervals. PC1 represents a continuum from lower elevations (positive loadings) to higher elevations (negative loadings). PC2 represents a continuum from browsing (negative loadings) to grazing (positive loadings). (From Schroeder et al. 2010)

areas used more rugged terrain at higher elevation than males (Schroeder et al. 2010).

Bighorn sheep exhibited diverse diets, as estimated from microhistological analyses of feces (Schroeder et al. 2010). Bighorn sheep in the Mono Basin ate mostly forbs during winter. Bighorn sheep in the Southern Region, which occurred at lower elevations, consumed mostly shrubs and graminoids. Indeed, principal component analysis (PCA) indicated that bighorn sheep in the Mono Basin exhibited differential patterns in diet composition compared with bighorn sheep in the Southern Region (Fig. 30). PC1 likely represented a continuum in altitude from lower elevations (positive loadings) to higher elevations (negative loadings). Spatial separation of sexes occurred in both the Mono Basin and the Southern Region. Additionally, more spatial (differential use of rugged terrain by the sexes) and less dietary separation between sexes was documented in the Mono Basin than in the Southern Region. Conversely, the sexes of bighorn sheep in the Southern Region exhibited less separation in use of rugged terrain, with considerably less overlap on the dietary niche axis than for bighorn sheep occupying the Mono Basin. Those forage niches differed most when sexes of bighorn sheep overlapped more in spatial distribution and differed less where spatial separation was more pronounced, emphasizing the need for a niche-based approach to understand ecological segregation of the sexes (Table 5). Thus, sexual segregation resulted from an integration of spatial distribution, diet, and terrain.

SUMMARY

This chapter provides an overview of organizing principles for understanding sexual segregation and focuses on methods of detecting and measuring sexual segregation. The chapter offers a broad treatment and examples of effects of scale on measurements related to sexual segregation, and it provides a synthetic approach for understanding sexual segregation in the context of a niche framework.

1. Hypotheses for sexual segregation must consider the adaptive significance of factors leading to separation of the sexes; purely mechanistic approaches are unlikely to provide adequate explanations.
2. Hypotheses for sexual segregation that are not linked with degree of sexual size dimorphism do not offer reasonable stand-alone explanations; among ungulates, it is primarily sexually dimorphic ruminants that sexual segregate.

3. Differences between proximate and ultimate causations of sexual segregation in ungulates should be assessed and divided into those that hold potential to explain this phenomenon across an array of ungulates and those that are specific to a particular species or environment. Those variables without wide applicability may contribute to the degree of sexual segregation under specific circumstances but will not offer a general explanation.

4. There must be seasonal variation in proposed measures for the degree of sexual segregation— the sexes must segregate and aggregate in an annual pattern, especially in seasonal environments.

5. The sex ratio must differ between areas or from the sex ratio of the population as a whole during sexual segregation.

6. Activity patterns must differ and be less synchronous for mixed-sex than single-sex groups. These patterns must be linked and integrated into other metrics of sexual segregation, such as use of food, habitat, or space.

7. Social factors thought to cause sexual segregation must be integrated into temporal changes in the use of food, habitat, or space.

8. Scale must be considered in measuring sexual segregation. The failure to detect sexual segregation may be the result of sampling at the wrong scale.

9. Space, habitat, and diet must be evaluated in concert to detect whether sexual segregation occurs.

10. Strengths and weaknesses of metrics for detecting sexual segregation are discussed, including segregation coefficients, the sexual segregation and aggregation statistic, tests of ratios and proportions, multi-response permutation procedures, and information-theoretic and machine-learning algorithms.

11. Effects of measurement scale are discussed relative a number of life-history characteristics, including population density, home ranges, habitat selection, and sexual segregation.

12. A synthetic approach of understanding sexual segregation is conceptualized with respect to the niche dynamics of the sexes related to space, habitat, and diet. Forage and habitat use differed most when sexes overlapped more in spatial distribution and differed least where spatial separation was more pronounced.

5 | Failed Concepts

Readers may wonder why I have waited so long to address topics considered in this and the next 2 chapters. In most scientific publications, hypotheses are offered at the end of the introduction. My reason for flouting this premise of scientific writing is that a substantial background was necessary to begin assessing hypotheses for sexual segregation. I was concerned that not everyone would have the contextual perspective necessary to follow my arguments. Moreover, I believed it would be more productive, and make my reasoning clearer, to review the underpinnings of hypotheses for potentially explaining sexual segregation as they relate to the ecology, morphology, physiology, and behavior of ungulates before proceeding to specific hypotheses for that phenomenon. With at least some of that background now in hand, I will tackle selected hypotheses forwarded to explain sexual segregation, starting with those least apt to explain this widespread phenomenon.

Setting Aside Rejected Hypotheses

Why do some hypotheses fall short as an explanation for sexual segregation? Hypotheses for sexual segregation may be subdivided into 2 broad categories: social and ecological (Mysterud 2000). Those categories are somewhat arbitrary, and hypotheses separated into those interrelated elements can never be truly independent. All interactions of animals with their environment—either social or ecological—ultimately are mediated by behavior. The key is to identify which aspects are primarily responsible for causing segregation of the sexes.

Social Hypotheses

One social hypothesis, forwarded by Bon and Campan (1996) and Bon et al. (2005), supports an ontogenetic (developmental) approach for understanding sexual segregation. This hypothesis involves a complex suite of causations and motivations for ungulates that are entirely proximate and mechanistic. The approach rests on the premise that sex-specific behaviors and sexual experience are important contributors to sexual segregation, and that social choices and sexual incompatibilities result in segregation of the sexes, sometimes referred to as the *social-preferences hypothesis*. Those explanations seldom indicate how such behaviors ultimately affect the reproductive success of individuals engaging in those behaviors (i.e., how or why those behaviors would be maintained by natural selection into adulthood). Indeed, few links are provided between development of behaviors in young and tests of how they are related to reproductive success of adults (where sexual segregation is prevalent among ungulates). For instance, Gaudin et al. (2015) reported that, in domestic sheep, young females were more strongly attached to their mothers than were young males, but evidence that this leads to sexual segregation of adults has not been demonstrated. Such approaches offer descriptions of sexual segregation (or its precursor) but lack a causal explanation. This hypothesis fails to explain why such behaviors would be adaptive in adults (i.e., favored by natural selection).

Bon and Campan (1996) argued that ontogeny is central because sexual segregation increases with age. Male ungulates do show variation in the degree of segregation with age (especially yearlings, which often remain in female groups). But such an age-related pattern is more poorly defined for females than males. Yet, in some species, it is the movement of females that produces sexual segregation (Bowyer 2004). Moreover, degree of segregation is not a smooth clinal process dependent solely on age and size. Sexual segregation is most pronounced for adult males and females (Bowyer 2004). Age and reproductive activity can be important determinants of spatial behavior of male and female ungulates (Malagnino et al. 2021). Ontogeny, however, provides a description of what happens but is not necessarily a satisfactory explanation for why sexual segregation in ungulates occurs.

Further, the seasonal pattern of sexual segregation and aggregation is variable (see Figs. 17–20) and difficult to explain with a developmental approach, which accounts poorly for annual changes in types and sizes of social groups. Hormones have been invoked to explain this pattern (Bon and Campan 1996), but many ungulates that do not sexually segregate also undergo similar hormonal changes. Likewise, simply invoking motivations related to parental care will not suffice to explain sexual segregation around the time of parturition, because many species with attentive mothers do not sexual segre-

gate (Clutton-Brock 1991). Perissodactyls ostensibly undergo many ontogenetic and hormonal changes similar to those of sexually dimorphic artiodactyls but do not exhibit sexual segregation.

Many of the predictions made by Bon and Campan (1996) and Bon et al. (2005) for ecological hypotheses for sexual segregation, and then forwarded to necessitate invoking an ontogenetical approach, were incorrect. For instance, the prediction that increased population density and reduced food availability would result in greater spatial segregation of the sexes has been falsified experimentally (Kie and Bowyer 1999) and will be discussed more fully in Chapter 7. Bon et al. (2005) contended that ecological factors alone could not account for the phenotypic variation related to body size, sex, age, or social status observed with respect to sexual segregation. Notably, those authors omitted the gastrocentric hypothesis (Barboza and Bowyer 2000, 2001; see Fig. 8), which holds potential to explain those outcomes related to age and body size. Recall that food and risk or predation are involved in group formation, which affects degree of polygyny, male-male combat, sexual dimorphism in size, and ultimately sexual segregation (see Fig. 4).

An oversimplified and narrow concept of what constitutes habitat in models forwarded to understand an ontogenetic approach for sexual segregation (Bon et al. 2005) renders models to examine social factors unrealistic. Moreover, that approach does not consider dietary differences between sexes. Although heterogeneous habitats can be a cause of segregation (Bowyer et al. 2002b), the assumption that heterogeneous habitats are necessary to cause sexual segregation is incorrect. Variation in the distribution of resources within a specific habitat can be sufficient to cause segregation of the sexes (Bowyer 1984, 2004). Moreover, the hypothetical models used to infer social causations of spatial differences between sexes have ignored effects of scale, which are pervasive in this process (Bowyer et al. 1996; Chapter 4). A more complete understanding of what constitutes habitat would help avoid misconceptions about the role of habitat in explaining sexual segregation, and would draw attention to the many facets that compose habitat (Hall et al. 1997)—a clear need exists for standardized definitions of habitat (Krausman and Morrison 2016). Too narrow a definition of habitat, for instance those excluding environmental attributes of the thermal landscape (Long et al. 2014), offers an oversimplified view and would affect conclusions concerning causes of sexual segregation.

The notion that sexual segregation does not occur in the absence of predators (Bon et al. 2005) is inaccurate. Although some antipredator behaviors of ungulates may wane in the absence of predation, others such as changes in group size and composition with respect to amount of concealment cover (Bowyer et al. 2001a) may persist for thousands of years (Berger et al. 2001c). Sexual segregation in red deer on the Scottish island of Rùm, where large carnivores are absent, has been well documented (Clutton-Brock et al. 1982a). LaGory et al. (1991) also reported sexual segregation in white-tailed deer on Ossabaw Island in the state of Georgia in the absence of predators. Likewise, McCullough (1979) reported sexual segregation in white-tailed deer on the George Reserve in southeastern Michigan, an area from which large carnivores were excluded. In addition, the supposition that animals might congregate from environmental conditions, but then disperse without some social attraction (Bon et al. 2005), is ingenuous and disregards nearly 5 decades of research on the evolution of group living (Chapter 2). For instance, a purely social explanation for gregariousness fails to account for variation in group size with changes in concealment cover and predation risk or for why multiple groups should form within seemingly homogenous patches of habitat (Bowyer et al. 2001a). Further, Bon et al. (2005) believe it is necessary to define components of segregation without referring to individual mechanisms. Sexual segregation is a population-level process, but it results from behavioral decisions by individuals. Natural selection operates principally at or

below the level of the individual organism; invoking group-level adaptations offers an unlikely evolutionary pathway for sexual segregation (Williams 1966). I am not discounting the importance of behavior in understanding gregariousness of ungulates. For example, group size plays a major role in the type of mating systems that evolve (Bowyer et al. 2020a). I question whether social mechanisms alone represent an overarching process sufficient to cause sexual segregation.

A serious problem with the ontogenetic hypothesis is that proposed causations are not separable from a number of ecological hypotheses (i.e., gastrointestinal morphology and physiology is altered with sex and age, as is predation risk). In addition, social hypotheses with their mechanistic underpinnings are entirely proximate and without an explanation for how they might be maintained by natural selection—how do such behaviors benefit the reproductive success of individual males and females?

If one ecological explanation for sexual segregation does not hold, it is poor logic to assume that the only alternative is a social hypothesis. Correlations and associations used to promote ontogenetical ideas may not represent causations. Indeed, the observed patterns could be caused by a number of ecological processes that typically have not been considered in studies of social segregation (Bowyer 2004). Worse yet, for the most part, this hypothesis is not testable. Nonetheless, one experimental study, using domestic sheep and small paddocks (1 ha), reported neither differences in activity of males and females nor changes in group composition (Michelena et al. 2004). Nevertheless, small-scale experiments may be satisfactory for testing some aspects of sexual segregation, but such artificial conditions will not likely suffice for understanding social hypotheses for free-ranging ungulates. Simply put, the ontogenetic approach offers a useful description of sexual segregation rather than providing an explanation for why it occurs. An additional problem with all of the social hypotheses is that, although some can potentially lead to the differential use of space by the sexes, none provide an independent answer as to why specific and different areas, sometimes spaced many kilometers apart, are used by the sexes during segregation. I believe that an ontogenetic approach offers little chance of adequately explaining sexual segregation and should be set aside to permit a stronger focus on other viable hypotheses (Chapter 7).

Another social hypothesis requires that males engage in altruistic behavior by leaving superior ranges to minimize competition with females and young (Geist and Petocz 1977, Geist 1982). For polygynous ungulates, mating opportunities are restricted mostly to large, dominant males that can restrict access of subordinate males to estrous females (Bowyer et al. 2020a; Chapter 2). The unlikely hypothesis of group selection (Williams 1966) must be invoked to explain why males that had not mated would voluntarily leave areas with superior forage, potentially compromising their own fitness, to avoid competition with unrelated young—an unlikely evolutionary outcome (Main and Coblentz 1990). Thus, subordinate males forgoing advantages by segregating into areas with reduced foraging opportunities to assist unrelated young can be set aside as a workable hypothesis for sexual segregation.

Verme (1988) postulated that male cervids segregate to open areas to allow visual assessment of other males to help establish dominance hierarchies (both difficult to test). He further proposed that use of those open areas prevented damage to growing antlers, a hypothesis first forwarded by Boner (1861). This hypothesis predicts that ungulates inhabiting open habitats should not sexually segregate (Main and Coblentz 1990), yet segregation occurs in numerous open-land ungulates (Bowyer 1984, Bleich et al. 1997, and many others). Recall that sexual segregation is near ubiquitous among sexually dimorphic ruminants (see Chapter 2 and Table 1), which this hypothesis fails to consider. Relating sexual segregation to protection of growing antlers is problematical because it offers a too narrow and too limited explanation for sexual segregation—and it cannot explain why species with horns sexually segregate (Main and Coblentz 1990). Further, this hypothesis does not explain why monomorphic ungulates, such as muntjacs (*Muntiacus* spp.) that live in forests and possess small antlers, do not sexually segregate.

Geist and Bromley (1978) proposed that sexual segregation was an antipredator strategy because males exhausted from rutting activities did not immediately shed their antlers, thereby indicating to potential predators that they were not vulnerable females. The subsequent shedding of antlers was interpreted as a form of female mimicry that enabled males to rejoin females to minimize the risk of being selectively preyed upon (Geist and Bromley 1978). This hypothesis fails to account for segregation occurring prior to rut (Bowyer 1984, McCullough et al. 1989), and it has limited empirical support (Main and Coblentz 1990). Moreover, this hypothesis cannot apply to ungulates that to do not shed antlers, such as bovids, and has little likelihood of explaining the widespread nature of sexual segregation.

Ecological Hypotheses

The Bell-Jarman Principle, which was formulated to explain dietary differences among species of ruminants, has been invoked as an explanation for sexual segregation between the sexes (Chapter 2; Demment 1982, Demment and Van Soest 1985, Pérez-Barbería et al. 2008). Variation in body size of ruminant species in relation to their gut capacity and metabolic rate is thought to result in the diversification of those ungulates into numerous foraging niches. Whether this hypothesis is applicable to differences between the sexes of sexually dimorphic ruminants, however, has been questioned (Chapter 2; Barboza and Bowyer 2000, Bowyer 2001a, Stewart et al. 2011). Weckerly (2010) documented that the scalar for the rumen-reticulum relationship with body size was not greater than the scalar for the metabolic relationship for the sexes of white-tailed deer. Hence, he rejected the Bell-Jarman Principle as a general explanation for sexual segregation by dimorphic male and female ruminants. This hypothesis also lacks power to explain seasonal patterns of variation in sexual segregation. Indeed, the Bell-Jarman Principle fails to address morphological changes in the digestive tract of females associated with late gestation and lactation and their subsequent effects on diet selection and processing (Barboza and Bowyer 2000, 2001), and it should not be conflated with the gastrocentric hypothesis. The Bell-Jarman Principle holds little promise to explain differences between the sexes.

McCullough (1999) and Bliss and Weckerly (2016) promoted the hypothesis of *female substitution* as an explanation for sexual segregation. This hypothesis rests on the assumption that a reduction in males creates capacity for an increase in number of females. Indeed, adult sex ratios skewed toward females are typical of dimorphic ungulates (McCullough 1999, Toïgo and Gaillard 2003). Because of sexual dimorphism in size and sexual differences in morphology and physiology of digestive systems, niche require ments of males are broader than those of females, and males can subsist on lower-quality but more abundant forage, which differs markedly from dietary needs of females (Chapter 2). Consequently, the broader foraging niche of males compared with females is proposed to reduce competition for resources between sexes and thereby increase carrying capacity (K) for females. This hypothesis is consistent with sexual size dimorphism being involved in sexual segregation in ungulates, and also conforms to some of what we understand concerning differences in digestive morphology between the sexes. Nonetheless, this hypothesis falls short in several areas. First, changes in degree of sexual segregation between high and moderated densities of white-tailed deer were not a result of skewed sex ratios (Kie and Bowyer 1999). Second, changes in the digestive tract of reproductive females were yet to be described when McCullough (1999) proposed the female-substitution hypothesis. Without such changes in female morphology and physiology, the female-substitution hypothesis lacks a mechanism to explain seasonal patterns of segregation and aggregation. Thus, the gastrocentric hypothesis (Barboza and Bowyer 2000, 2001; see Fig. 8) supersedes the hypothesis of female substitution as an overarching hypothesis for sexual segregation.

The related *sexual dimorphism–body size hypothesis* of Main et al. (1996) invokes physiological factors connected with nutrition as being responsible for sexual segregation, with each sex using resources that meet their different physiological requirements. This premise is too broad to easily test and fails to make explicit testable predictions (Bowyer 2004). This hypothesis can be subsumed under the gastrocentric hypothesis.

Clutton-Brock et al. (1987) experimentally tested whether males avoiding parasites in the feces of red deer would produce sexual segregation. Males might seek out areas with fewer parasites (indexed by the amount of feces that harbor such parasites) than would females because females have other constraints related to gestation and lactation. Clutton-Brock et al. (1987) reported no evidence that males investigated or avoided high densities of feces from either sex, making this hypothesis unlikely. Nonetheless, Ferrari et al. (2010) hypothesized that if males possessed a higher parasite burden than females, and if males were primarily responsible for transmission of parasites, females would benefit from not frequenting highly infested areas used by males, thereby reducing the parasite infection in females. Those authors used models for alpine ibex to simulate effects of host-parasite interactions to demonstrate different degrees of sexual segregation, but no data exist to support this assertion. This hypothesis does not adequately address why there are seasonal patterns in sexual segregation (Chapter 3). Moreover, the hypothesis fails to explain why male bighorn sheep move to and from female ranges to produce patterns of aggregation and segregation (Bleich et al. 1997). If parasite transmission between sexes produces spatial patterns on the landscape, this arrangement is more likely a function of other factors promoting sexual segregation, rather than parasites being the cause.

Zuk and McKean (1996) reported that male nyala (*Tragelaphus angasii*) were more heavily infected with arthropod parasites than females. Those differences in parasite burdens were hypothesized to be the result of sexual differences in behavior, ecology, or physiology. Nonetheless, such sexual differences were likely an outcome rather than a cause of sexual segregation.

Rodgers et al. (2021) examined differences between the sexes in fidelity, timing, distance traveled, and *green-wave surfing* (following patterns of vegetation green-up in spring) during migration of mule deer as explanations for sexual segregation. They did not, however, provide evidence that the sexes used or selected habitats differently, or provide evidence on potential dietary difference between the sexes. Also, those authors proposed to test the reproductive-strategy and forage-selection hypotheses, which are both difficult if not impossible to test (see later in this chapter). Moreover, many mule deer that do not migrate do sexually segregate (Bowyer 1984). In addition, the sexes of zebra (*Equus burchelli*) make extensive movements (Bartlam-Brooks et al. 2011) but do not sexually segregate. If these migratory patterns could be shown to involve sexual segregation, they likely would be subsumed under other ecological explanations for sexual segregation, such as forage or habitat (Chapter 7). Clearly, migratory patterns do not provide a stand-alone hypothesis for sexual segregation and should be set aside as a viable hypothesis.

Limiting Overutilization of Habitats

In an otherwise valuable contribution on sexual segregation in African elephants, Kioko et al. (2020) proposed that sexual segregation ultimately aided in distributing elephant foraging pressure across different habitats, which served to limit overutilization of specific habitats by elephants. Ungulates may improve habitat quality and diversity of plants via grazing and deposition urine and feces (McNaughton 1979b; Stewart et al. 2006, 2009), and sexual segregation affects the distribution of ungulates on the landscape (Bowyer 2004). The limitation of overutilization, however, implies either self-regulation of animal populations (McCullough 1979) or altruistic behavior to benefit the species or environment (Williams 1966)—neither outcome is a likely result of sexual segregation.

Problematical Speculations

Hypotheses for sexual segregation can be problematic for several reasons. Chief among these is that some are not independent of one another and, hence, do not offer distinct explanations for the widespread nature of sexual segregation (i.e., the predictions for different hypotheses overlap or are identical). Others may not be true hypotheses for *why* sexual segregation occurs but instead are predictions or are observations of *what* has occurred. A few are too proximate to offer an explanation for the widespread occurrence of sexual segregation (e.g., males avoiding damage to antlers) and relate primarily to individual species and site-specific conditions, which might correlate with or augment more ultimate causations for sexual segregation.

Reproductive-Strategy Hypothesis

The *reproductive-strategy hypothesis* predicts that ecological factors and social behaviors interact to cause sexual segregation (Main et al. 1996, Main and du Toit 2005, Main 2008). Because reproductive success of males is influenced by size, strength, and fighting ability, natural selection favors behaviors that maximize rate of growth and formation of energy reserves. Those reserves, in turn, enhance the ability of males to compete for females and thereby increase their reproductive success. Reproductive success of females, however, is determined primarily by survival of offspring; thus, females engage in activities that provide resources necessary to meet or exceed requirements for gestation and lactation to promote their own survival and that of their young. Consequently, predation risk often is postulated as a key component of this hypothesis. DePerno et al. (2003), however, treated predation risk and the reproductive-strategy hypothesis as separate entities, which I believe is an important distinction.

One problem inherent to the reproductive-strategy hypothesis (Main et al. 1996) is that this hypothesis is too far-reaching to be testable. Many characteristics of ungulates can be attributed to some reproductive strategy (Bowyer 2004). Further, Bowyer (2004) and Neuhaus et al. (2005) observed that the reproductive-strategy hypothesis was cast too broadly to provide specific predictions about sexual segregation. Other hypotheses may explain many adaptations of ungulates related to their behavior, ecology, or physiology. Whether such interpretations are associated with causes of sexual segregation or simply related to other aspects of ungulate biology is uncertain and difficult to unravel. Nevertheless, risk of predation, which I believe is a viable stand-alone hypothesis, is testable. Most problematical, however, is that many hypotheses (especially the reproductive-strategy hypothesis) forwarded to explain sexual segregation are not independent of one another and, consequently, are not testable. In any event, ambiguities in the formulation of the reproductive-strategy hypothesis make its value in understanding sexual segregation nil (Stewart et al. 2011).

Forage-Selection Hypothesis

Tests of the forage-selection hypothesis (Main et al. 1996, Main 2008 for reviews) have become widespread in the literature on sexual segregation (Stewart et al. 2011). This outcome is disappointing because the forage-selection hypothesis is not a hypothesis but, at best, is a prediction or observation. This "hypothesis" provides no reason for why the sexes should aggregate or segregate, it simply notes that they forage differently. That the sexes forage differently requires an

explanation (i.e., hypothesis) for why they should do so. For sexually dimorphic ruminants, only one viable hypothesis for sexual segregation holds potential to explain differences in morphology and physiology between sexes that would lead to disparities in foraging—the gastrocentric hypothesis (Barboza and Bowyer 2000, 2001; see Fig. 8). Several authors have conflated differences between the gastrocentric and forage-selection hypotheses and their predictions by subsuming this true hypothesis (gastrocentric) under the mantel of the forage-selection hypothesis (Ruckstuhl 2007, Main 2008, Gregory et al. 2009), which provides no explanation for why the sexes should forage differently. Regrettably, once an idea becomes ingrained in the literature, it is extraordinarily difficult to eliminate no matter how inadequate or unworkable it might be. Considerable thought and effort on the part of those studying sexual segregation will be required, but the forage-selection hypothesis needs to be eliminated as an explanation for sexual segregation.

In sum, hypotheses postulated by Main et al. (1996) suffer from a lack independence. For example, differences between sexes proposed for explaining the sexual dimorphism–body size hypothesis also relate to the differential susceptibility of males and females to predation (Pierce et al. 2000a, b), which is incorporated under the reproductive-strategy hypothesis. Social factors, especially aggressive interactions, ostensibly are related to sexual dimorphism in body size, and assorted reproductive strategies almost certainly involve social behavior. The need to organize hypotheses for sexual segregation into a testable framework is urgent.

Proximate Causations

Several hypotheses forwarded to explain sexual segregation are proximate and extremely mechanistic in nature; these relate to individual species and site-specific conditions rather than offering a cause for the widespread occurrence of sexual segregation in dimorphic ungulates. Difficulties associated with antler sensitivity in providing a general explanation for sexual segregation were addressed previously. The most common proximate hypothesis for sexual segregation, however, concerns weather sensitivity of some species under particular conditions. Staines (1977) reported that red deer sought shelter from chilling winds during winter by using more sheltered areas, and that the sexes differed in their use of such areas. Conradt et al. (2000) further noted that wind speed was lower at sites occupied by males than those used by females on windy days. Large males were thought to be more sensitive to cold weather because their net energy gain decreases with absolute heat loss more rapidly than that of smaller females, thereby providing a mechanism for sexual segregation (Conradt et al. 2000). Ungulates have an extensive geographic distribution (Chapter 1), and many species are not confronted by the winter conditions experienced by red deer in Scotland, yet sexual segregation is widespread (Bowyer 2004; see Table 1). Other temperate ungulates sexually segregate during winter (McCullough 1979, Shank 1982, Miquelle et al. 1992, Brockmann and Pletscher 1993, Bowyer et al. 2001b), but sexual segregation in many species is most prevalent during and following parturition (Bowyer 2004). North American elk are more efficient at dissipating heat at high temperatures than mule deer (Parker and Robbins 1984), yet both species exhibit sexual segregation. Thus, weather sensitivity is likely to augment or interact with other factors to further promote segregation of the sexes under particular conditions. I do not discount observations concerning winter severity, but I believe they offer a proximate rather than an ultimate explanation for why the sexes segregate. I view weather sensitivity as a component of habitat selection that needs to incorporate a complete set of weather variables in addition to temperature (Bowyer and Kie 2009, Long et al. 2014) and other features of habitat (Hall et al. 1997). Weather sensitivity is not a viable stand-alone hypothesis. Finally, weather sensitivity likely is experienced by perissodactyls that do not sexually segregate.

Similarly, Cohen et al. (2021) proposed that sexual segregation was driven by inherent preferences of endotherms for differences in temperature, using migratory eastern Mediterranean birds and bats as evidence for this proposition. Several difficulties exist with this proximate approach, including extending their arguments from a geographically limited area for migratory birds and bats to all endotherms. Indeed, not all species that sexually segregate are migratory, which further limits the scope of their conclusions. Moreover, differences in wing loading related to migration between sexes of some birds (sandpipers, *Calidris* spp.) is thought to be associated with differential predation risk and the escape performance of the sexes (Burns and Ydenberg 2002). Likewise, wing loading in little brown bats (*Myotis lucifugus*) affects habitat use via changes in maneuverability during foraging in less cluttered areas (Kalcounis and Brigham 1995), rather than just migration. Temperature may affect physiological responses of the sexes, but it also alters plant communities, prey availability, and risk of predation. Numerous advantages for migration exist that are related to more variables than simply temperature (Fryxell and Sinclair 1988, Avgar et al. 2014). Finally, there is no direct explanation (or evidence) for how differences in temperature preferences in birds and bats affect reproductive success—as with other temperature-sensitivity hypotheses, the approach is entirely too proximate.

Several authors have reported that female ungulates move to areas with free water or remain closer to water sources than males during sexual segregation, particularly around the time of parturition (Bowyer 1984, Main and Coblentz 1996, Bleich et al. 1997). Larger-bodied mammals have lower rates of water loss than smaller-bodied ones (Gordon 1977). Lactating ruminants have higher water requirements than males or nonlactating females (Short 1981, Barboza et al. 2009). Consequently, there is a compelling reason for lactating females to be closer to water sources than males. Indeed, water can play a major role in the ecology of some ungulates (Bleich et al. 1997, Whiting et al. 2012, Heffelfinger et al. 2020).

Nonetheless, whether the distribution of free water can explain sexual segregation across a wide array of ungulates is questionable (Bowyer 1984). Again, water is likely a habitat component that augments other factors largely responsible for sexual segregation on a more widespread basis.

Avoiding More Speculation

Experimental research allows a more careful method of interpreting cause and effect than either observational or correlative studies. A hypothetico-deductive design provides a rigorous approach for testing predictions, including those for sexual segregation—the strength of that method is that it allows for falsification of hypotheses (Popper 1959). Errors in reasoning may result from observational or correlational studies that rely solely on *inductive reasoning* (i.e., moving from specific observations to broad generalizations). Such studies can commit Popper's (1959) white-swan fallacy: no number of observations of white swans can confirm the hypothesis that all swans are white because the observation of a single black swan falsifies that presumption. Critical tests of properly framed hypotheses will answer questions about *why* particular events occurred and hold promise to be generalizable, thereby leading to a greater understanding of processes responsible for sexual segregation. Less-general tests, focused on questions about *what* happened, predictably will be narrow in scope, and often rely on correlations or observational data alone. Such correlational results run the risk of confusing cause and effect, especially when there are lurking variables. Conducting experimental research on free-ranging ungulates can be a daunting task (McCullough 1979, Stewart et al. 2006), but results from such studies offer far greater certainty of outcomes than those from correlational studies. Moreover, this approach will help reduce unwarranted speculation.

What is necessary to formulate hypotheses to obtain critical tests of sexual segregation in ungulates? First, such tests must relate to sexual dimorphism in ruminants—sexual segregation occurs primarily among this group of ungulates (Chapter 2). Hypotheses that apply equally well to other ungulates that are not sexually dimorphic in size will not suffice as ultimate causations for sexual segregation, because those animals do not sexually segregate. For example, a hypothesis invoking hormones or maternal care as causing temporal variation in degree of sexual segregation is not a sufficient stand-alone hypothesis because other species that do not sexually segregate possess similar physiological and behavioral characteristics.

Second, hypotheses must cope with the temporal and spatial nature of sexual segregation, including the differential use of particular spaces (e.g., areas used for parturition or rut), habitats, and forages used by the sexes. Information on distances between sexes during periods of segregation and aggregation are necessary to fully evaluate such hypotheses, a result that often is omitted. Moreover, the spatial scale at which sexual segregation occurs must be considered. Varying scales of measurement can have a profound effect upon the results of such research (Bowyer et al. 1996, Bowyer and Kie 2006; Chapter 4). Care must be taken to assure that scale is not biasing conclusions regarding the biology of the animals being investigated. Indeed, in some instances, it is difficult to ascertain whether a potential outcome related to sexual segregation is a result of biology or sampling bias (Bowyer 2004).

Third, care should be taken to assure that hypotheses proposed for sexual segregation lead to predictions about *why* an outcome occurs. Predictions that simply indicate what is happening will not resolve why it is occurring (e.g., the forage-selection "hypothesis"). This confusion of predictions (or observations) with hypotheses can obscure the actual causes of sexual segregation and divert efforts that might otherwise have been more productive (Stewart et al. 2011). The most useful hypotheses must be able to distinguish cause and effect.

Fourth, the forwarding of hypotheses that are not independent should be avoided. If aspects of the ecology and behavior are explained equally well by multiple hypotheses for sexual segregation, those hypotheses are not independent and will be difficult to differentiate from one another (e.g., reproductive-strategy hypothesis). This is conceptually different from hypotheses that are not mutually exclusive. When hypotheses are not mutually exclusive, one hypothesis or the other may cause sexual segregation, or both may operate to do so. Under particular conditions, one factor may override effects of the other, thereby increasing complexity. The key is that each must be amenable to a critical test of whether it promotes sexual segregation.

Fifth, I argue that this broad occurrence across taxonomically and morphologically different species requires a more general (i.e., ultimate) explanation for sexual segregation. Indeed, I believe that sexual segregation meets Williams' (1966) onerous definition for an adaptation brought about via natural selection (Bowyer 2004). Narrow mechanistic explanations for sexual segregation will not suffice to explain this phenomenon. Moreover, limiting hypotheses for sexual segregation to proximate conditions (e.g., weather sensitivity, availability of free water) will not provide a sufficiently broad premise for understanding why the sexes segregate.

Sixth, there is much to be learned from a comparative approach. Nevertheless, the lack of a universally accepted definition for sexual segregation and the myriad of hypotheses forwarded to explain this phenomenon makes a comparative approach futile (Bowyer 2004). Indeed, there are few studies where all of the necessary data to test both social and ecological segregation have been collected for ungulates. The only consistency for sexual segregation in ungulates is that it is strongly related to sexually dimorphism in ruminants. Moreover, some good ungulate analogues for studying sexual segregation exist (e.g., elephants, kangaroos, and their relatives), but other mammalian groups (especially carnivores) are of limited value in this regard. For instance, infanticide in grizzly and brown bears (*Ursus arctos*) and black bears (*U. americanus*), which is thought to affect sexual segregation in those large

carnivores (Wielgus and Bunnell 1995, Rode et al. 2006, Gantchoff et al. 2019), offers an improbable explanation for large, herbivorous mammals that seldom engage in that behavior (Bowyer 2004). Too often hypotheses developed for ungulates are applied to other taxa, and vice versa, with sometimes confusing results. There is a desperate need to relate the hypotheses for sexual segregation to the life-history characteristics of the species under study. I will elaborate more on what constitutes reliable comparisons among taxa of mammals in the following chapters.

SUMMARY

This chapter surveys a number of hypotheses that hold little promise for explaining sexual segregation in ungulates. The problematic properties of hypotheses are explained, and criteria for obtaining a critical of hypotheses for sexual segregation are delineated.

1. Hypotheses for sexual segregation may be subdivided into 2 broad categories: social and ecological.
2. Several social hypotheses should be set aside because of their failure to provide a meaningful explanation for sexual segregation. The ontogenetic (developmental) hypothesis, along with other difficulties, provides a description of sexual segregation rather than an explanation for its cause and is difficult to test empirically.
3. Another social hypothesis requires that males engage in altruistic behavior by leaving superior ranges to minimize competition with females and young. Subordinate males forgoing advantages by segregating into areas with reduced foraging opportunities to assist unrelated young would require invoking group selection, which offers an unlikely evolutionary pathway for sexual segregation.
4. Male cervids were hypothesized to segregate to open areas to allow visual assessment of other

males to help establish dominance hierarchies, and use those open areas to prevent damage to growing antlers. This hypothesis ignores that species with horns also exhibit sexual segregation, as do many open-land ungulates and, consequently, cannot provide a general explanation for sexual segregation. The hypothesis that sexual segregation is related to antler casting has this same limitation—not all ungulates that sexual segregate possess antlers.

5. Several ecological hypotheses are also problematic. The Bell-Jarman Principle, which has been proposed to explain niche differences among species of ungulates, also has been forwarded as the cause for sexual segregation within species. This hypothesis cannot explain seasonal differences in degree of sexual segregation, and differences between the sexes do not conform to predictions concerning their gastrointestinal tracts. Additionally, differential parasite loads between males and females offer an unlikely cause of sexual segregation.

6. The female-substitution hypothesis relies on digestive-tract differences between sexes to explain sexual segregation, but this hypothesis fails to explain seasonal differences in the degree of sexual segregation, and it has been superseded by the gastrocentric hypothesis.

7. Other hypotheses have inherent flaws in their construction. The reproductive-strategy hypothesis proposes that males and females possess divergent behaviors that result in them following different patterns for successful reproduction. This hypothesis is too far-reaching in scope to test. Many characteristics of ungulates can be attributed to some form of reproductive strategy; the reproductive-strategy hypothesis was cast too broadly to provide specific predictions about sexual segregation to be useful.

8. The forage-selection hypothesis is not a hypothesis but instead a prediction or observation (i.e., the sexes forage differently). It fails to explain *why* the sexes should forage differently.

9. Some hypotheses are far too proximate to offer a general explanation for the evolution of sexual segregation. Sensitivity to weather and needs for free water fall into this category. These factors likely augment or interact with other conditions that are responsible for sexual segregation and further promote segregation of the sexes under particular circumstances.

10. Several characteristics are related to reliable tests of sexual segregation. An experimental approach using a hypothetico-deductive test is superior to studies that relay on correlations, which may confuse cause and effect.

11. Reliable tests for sexual segregation must be related to sexual dimorphism in ruminants, because those species sexually segregate.

12. Tests for sexual segregation must cope with temporal and spatial differences in this process.

13. Hypotheses potentially explaining sexual segregation should lead to predictions about why an outcome occurs and be fully testable. Predictions that simply indicate what is happening will not resolve why it is occurring.

14. Hypotheses that are not independent will do little to resolve the causes of sexual segregation because such hypotheses cannot be differentiated from one another.

15. Hypotheses must not be so proximate or site specific that they fail to explain the evolution and near ubiquitous occurrence of sexual segregation in dimorphic ruminants.

16. Lack of an accepted definition for sexual segregation and the myriad of hypotheses forwarded to explain this phenomenon make a comparative approach across species futile.

6 | The Role of Social Behavior

Which behavioral hypotheses for sexual segregation have gained the most recognition in the literature? Herein I provide an in-depth evaluation of the 2 most prominent hypotheses for social behavior causing sexual segregation. These hypotheses are of markedly different origin, although obviously both ultimately relate to the social behavior of ungulates. Potential limitations exist to these explanations, but both hypotheses have been forwarded repeatedly in the literature as the cause of sexual segregation and, consequently, deserve special attention.

Social Aggression and Sexual Affinities

The first category of hypotheses ascribes either aggressive interactions or social affinities between the sexes outside rut as the cause of sexual segregation. These hypotheses typically involve males having an affinity for or a repulsion for one another, or avoiding or displacing the opposite sex, especially females avoiding males (MacFarlane and Coulson 2009). In one instance, sexual segregation is thought to occur because males avoid females to reduce rates of male-male aggressive interactions to increase time for other activities, such as foraging. Hence, Morgantini and Hudson (1981) postulated that minimizing sexually motivated aggression among males during pe-

riods when reproduction was not possible resulted in spatial separation of the sexes. Testosterone levels, however, decline following rut and in cervids are related to antler casting (Mirarchi et al. 1978, Demarais and Strickland 2011), factors which should lower aggression between males even if in association with females. Further, avoidance of such male-male aggression during the nonreproductive period lacks documentation (Main and Coblentz 1990). Moreover, this reasoning predicts that most polygynous ungulates should sexually segregate, but they do not do so.

The hypothesis for males avoiding females was attributed to Prins (1989) and Kie and Bowyer (1999) by MacFarlane and Coulson (2009). Large male African buffalo leave mixed-sex groups and form male-only groups, where they obtain better forage to help replenish body reserves (Prins 1989). Importantly, this behavior occurs over an extended mating season where males become exhausted from rutting behavior (Hay et al. 2008), and it would not apply to periods outside rut, when sexual segregation is prominent in temperate ungulates. Kie and Bowyer (1999) clearly rejected social factors as being responsible for sexual segregation in white-tailed deer; effects of social factors on sexual segregation should have increased at high compared with low density of deer, but the result was just the opposite. Indeed, males separating from females was explained primarily as an outcome from avoiding competitive interactions (Kie and Bowyer 1999). It is unclear how MacFarlane and Coulson (2009) arrived at their conclusion because neither publication they cited invoked social behavior to explain patterns of group composition.

In gray kangaroos, a good ungulate analogue, MacFarlane and Coulson (2009) suggested males possessed a strong behavioral affinity to maintain contact with one another and thereby contributed to the cohesion of male-only groups that promoted sexual segregation. The authors, however, do not provide quantitative data indicating benefits associated with males joining male-only groups. This hypothesis provides a potential (but I believe unlikely) explanation for why groups should separate, but not why they should aggregate or use specific areas during segregation. In addition, if males or females are avoiding one another, hypotheses related to social affinities cannot explain why mixed-sex groups are larger than single-sex groups, even outside of the mating season (Bleich et al. 1997, Kie and Bowyer 1999, Bowyer 2001a, and many others). Moreover, Weckerly et al. (2020) reported that male-male aggression limited, rather than promoted, the size of male-only groups of Roosevelt elk.

Females are typically thought to avoid males because sexual dimorphism in size results in adult males being socially dominant to females, a pattern that is common in gregarious ruminants (Weckerly et al. 2004, Peterson and Weckerly 2017). Other factors that do not require difference in body size may promote more intersexual aggressiveness in males than females. Male mammals may be dominant to females because males generally are the more aggressive sex—males typically compete more intensively for mates than females, especially in polygynous species. Male mammals may live in groups of unrelated individuals (Blundell et al. 2004), which promotes more aggression than in females, which are more likely to occur in groups with relatives (Jones

2014). Both of those factors are common to many mammals that do not sexually segregate. Nonetheless, adult male ungulates display more aggression than either females or younger individuals, winning most aggressive interactions with other sex and age classes (Koutnik 1981, Bowyer 1986a). Consequently, sexual segregation may be influenced by females avoiding aggression by adult male ungulates (Weckerly et al. 2001, 2004; Richardson and Weckerly 2007), but what benefits accrue to individuals that do so?

Aggressive interactions between sexes might adversely affect foraging behavior—potential costs to foraging females being interrupted by the presence of males ostensibly could be offset by moving away from those males, thereby creating spatial separation of the sexes. Costs related to aggressive interactions between males and females, however, seldom have been quantified. Nevertheless, Weckerly et al. (2001) observed female groups of Roosevelt elk moving away when approached by large groups of males. This observation, however, does not explain why the sexes use distinct areas at specific times during periods of segregation, or why extreme aggression by males during rut fails to produce similar patterns of spatial separation of the sexes. If intersexual aggression is a primary driver of sexual segregation, why do mixed-sex groups form both during and outside the rut (see Figs. 17–20)?

Weckerly (2001, 2020) argued convincingly that aggression limited the size of male groups of Roosevelt elk, while at the same time noting that ecological factors likely influenced the occurrence of males in groups composed largely of females. These results offer compelling arguments for constraints on the evolution of group living (Alexander et al. 1974, Krause and Ruxton 2002) but do not necessarily provide a complete and overarching explanation for sexual segregation. Moreover, the proposition that ecological segregation is more prevalent at larger scales and social segregation at smaller ones (Peterson and Weckerly 2017) fails to provide a link across scales or entirely explain the explicit distribution of the sexes on the landscape (in this instance, an ecological explanation also is required). Why sexual segregation could not occur at fine scales is unclear. Certainly, white-tailed deer sexually segregate at extremely fine scales (McCullough 1979, Kie and Bowyer 1999, Stewart et al. 2003b). Moreover, aggressive interactions and avoidance of males by females alone will not explain spatial patterns of sexual aggregation. Males behave most aggressively during rut when sexual aggregation is widespread—if there is a cost to associating with aggressive males, it must be offset by other factors. I believe the major contribution of this body of research is to provide a broader understanding of limits to sociality (i.e., group living) rather than providing an explanation for sexual segregation. Thus, male aggression may offer a limitation to the size of some social groups, but it falls short as an explanation for sexual segregation. This is a subtle but important point to which I will return later.

Activity Patterns

Calhim et al. (2006) argued that ecological hypotheses have focused almost exclusively on the occurrence of sexual segregation and have completely ignored the behavioral processes that lead up to segregation (e.g., fission and fusion of foraging groups and the levels of behavioral synchrony that preceded those events). This argument rests on 2 assumptions: factors related to behavioral synchrony are a cause rather than consequence of sexual segregation, and habitat and dietary preferences, as well as fear of predators, do not engender behavioral responses (an extremely narrow view given the vast literature on this topic). Nonetheless, the activity-budget hypothesis remains as a major explanation for sexual segregation in the literature.

The activity-budget hypothesis is based on differences in sexual dimorphism between the sexes of ungulates motivating changes in their activity patterns (Conradt 1998a, Ruckstuhl 1998). Body-size differences between males and females are thought to result in disparities in time spent feeding and moving, compared with bedding and ruminating, and

are based principally on differences in digestive morphology and physiology (Ruckstuhl 1998). Proponents argue that synchronization of activity and remaining in mixed-sex groups could be detrimental if optimal activity budgets and associated foraging patterns could not be achieved (Conradt 1998a, Ruckstuhl 1999, 2007). Sexual differences in activities might cause mixed-sex groups to fission and later fuse into single-sex groups. Hence, group fragmentation would result from sexual mismatches in activity budgets or costs of behavioral synchrony. Those social factors, then, would cause sexual segregation. The hypothesis, however, contains several paradoxes. First, it offers a mechanism for mixed-sex groups to fragment but not for them to merge. Second, if mixed-sex groups are the most prone to fragmentation, why are they larger than single-sex social groups? Clearly, factors other than differences in activity patterns of the sexes must be involved.

As an aside, many studies of ungulate activity patterns involve field measurements that sometimes do not include the complete diel cycle. Under such circumstances, calculation of neither an activity budget nor ensuing energetic costs is possible. Instead, it is often activity patterns during a period of daylight that are sampled—at least for some studies, the activity-budget hypothesis is inappropriately named. More importantly, drawing conclusions concerning activities or energetic costs without including the complete diel cycle can be misleading, especially when summed hours of activity or other behaviors are used as evidence for sexual segregation. For instance, Ruckstuhl (1998) noted that most activity in Rocky Mountain bighorn sheep (*O. c. canadensis*) was diurnal and, consequently, that those activities could be extrapolated to a diel activity budget. Nonetheless, another study of a desert-dwelling subspecies of bighorn sheep (*O. c. mexicana*) demonstrated bouts of nocturnal behavior (Alderman et al. 1989)—a pattern also present in more temperate species of ungulates (Beier and McCullough 1990). I believe it more appropriate to term this the "activity-pattern hypothesis," which might still support a general premise related to sexual segregation.

Why are there such differences of opinion concerning the validity of this hypothesis? Aside from the absence of an accepted definition for sexual segregation (Chapter 4), the activity-budget hypothesis is difficult if not impossible to test experimentally, and it must rely on observational and correlative studies as its foundation. Without experimentation, discerning cause from effect can be difficult. Moreover, there are limited data demonstrating that splitting of mixed-sex groups occurs as a result of intersexual differences in activity patterns. Many studies of activity patterns simply note differences in activity between the sexes in the various types of social groups and assume that they are measuring social segregation. Such assumptions rely on the notion that differences in activity between mixed- and single-sex groups reflect sexual segregation, rather than documenting how fusion and fission of groups affect activity patterns that might lead to sexual segregation—a critical point. Few studies measure the selective costs of occurring in mixed- versus single-sex groups or deal with potentially confounding effects of group size (Bowyer and Kie 2004). Moreover, differences in activity may simply be a correlate of sexual dimorphism and have little to do with sexual segregation (Bowyer 2004). As noted previously, some studies of social segregation are fraught with methodological problems.

A few attempts have been made to sort out some of the issues I raised with respect to shortcomings the activity-pattern hypothesis. Calhim et al. (2006) studied a population of feral goats (*Capra hircus*) to examine group fusion and fission related to several factors—principally, animals of different sexes joining or leaving groups. Those authors reported that the probability of fission increased with group size—unstable groups were larger than more stable smaller ones, but group composition did not affect the fission of groups independent of group size. Calhim et al. (2006) proposed that this resulted, in part, from there being too few small, mixed-sex groups, and they sought to correct this difficulty by evaluating mid-sized, mixed-sex groups of 4–10 individuals.

They concluded that mixed-sex groups were more likely to fission than single-sex groups (4 of 9 versus 5 of 16, respectively) and noted this was a significant outcome ($P < 0.01$). Their chi-square calculation, however, was incorrect, and group size remained a confounding variable in their analysis that was not adequately controlled—the reported dissimilarities were not significantly different.

Other problems existed with the study by Calhim et al. (2006). The authors did not consider solitary individuals in their analyses, introducing a potential bias relating to sexual segregation (Chapter 4). Moreover, they inflated sample sizes (which is more likely to produce a significant result) by sampling the same individuals repeatedly while using statistical procedures that required sampling without replacement. This statistical problem is not unique to Calhim et al. (2006), but others at least have tried to correct for this shortcoming by reducing alpha (the value necessary to detect a significant outcome) (Chapter 4).

Before further evaluating hypotheses for social segregation, I need to provide information on several other difficulties with the activity-patterns hypothesis. The first relates to calculating an energy budget for males and females. I already have addressed complications from not calculating activities for a complete diel cycle. Clearly, obtaining data on groups size and composition and on activity of individuals throughout a diel cycle is challenging but necessary. Perhaps more perplexing, however, is the definition used for active and inactive individuals employed to assess energetic costs from differences between the sexes in activity. Ruckstuhl (1998) and Ruckstuhl and Neuhaus (2005), and an extensive number of other publications following their lead, have identified bedding and standing as inactive states, and walking and foraging as active. Robbins (1983), however, reported that the cost of standing averaged 24.8% higher than the cost of bedding for several taxa of mammals, and Fancy and White (1985) noted for ungulates that those costs were 10% to 35% higher—standing is obviously a more costly activity than bedding and should not be considered as inactive.

This discrepancy raises important questions concerning the reliability of this approach.

In my experience, transitions between various states of activity in ungulates often are fluid, with one type of activity integrating quickly into another, sometimes with a return to the previous activity. There are a number of behavioral states of activity other than just walking and foraging, including alert postures and other predator-avoidance behaviors (see Table 2). Gates and Hudson (1978) reported that slow-walking had only a slightly higher energetic costs than standing, indicating the cost of those behaviors were similar. Fancy and White (1985) noted, however, that increasingly rapid gaits of ungulates were accompanied by rising energy expenditures. Clearly, other states of activity, including those used to elude predators, have considerably higher energetic costs that just walking or foraging, and they certainly should be included in calculating energy budgets.

Ruckstuhl (1998, 1999) proposed that differences in activity between sexes could influence their spatial separation, and that these differences were the ultimate explanation for all types of sexual segregation, including ecological hypotheses; causations other than activity patterns were thought to be secondary factors (Ruckstuhl 2007). Those mechanistic hypotheses have been extended and elaborated as explanations for why the sexes engage in all types of sexual segregation (Ruckstuhl and Neuhaus 2000, 2005), including a modeling effort by Ruckstuhl and Kokko (2002). A more complex model (Yearsley and Pérez-Barbería 2005), however, failed to find support for the activity-budget hypothesis. Problems associated with sampling of and definitions for activity, and with obtaining experimental tests of the activity-pattern hypothesis, also make this explanation for sexual segregation in ungulates unlikely.

A Case Study of Mule Deer

Research on mule deer on East Mesa in the Cuyamaca Mountains of southern California by Bowyer and Kie

(2004) assessed effects of activity patterns on sexual segregation. Those authors reasoned that if the costs from foraging in mixed-sex groups was greater than that for single-sex groups, that particular constraint on grouping would be revealed in differences in foraging efficiency (percentage of active deer feeding; Berger 1978) in mixed-sex groups compared with single-sex groups. Hence, if foraging efficiency (a near instantaneous measure of cost or benefit) was related to sexual segregation, foraging efficiency should be lower for mixed-sex than male-only or female-only groups throughout the year. Similarly, foraging efficiency of mixed-sex groups should be lower during periods sexual segregation than during sexual aggregation if foraging efficiency is the ultimate reason for sexual segregation. Moreover, if differences in foraging efficiency are a cause rather than a consequence of sexual segregation, foraging efficiency should be most affected by the type of social group (e.g., mixed sex, male only, female only) in which animals occurred compared with other components of sociality, such as group size.

The initial analysis was to determine whether males and females spatially separated from each other during one season and then aggregated in another, although this separation of sexes is seldom complete at any time of year (see Figs. 19, 20). In this Mediterranean ecosystem, there were 2 seasons: dry (May–October) and wet (November–April). The dry season encompassed parturition (June and July) and the wet season included rut (November and December). The study area was composed of a series of meadows preferred by deer that averaged ~78 ha but were separated by other vegetation types typical of montane southern California (Bowyer 1984, 1986b). Eastern meadows were quite dry compared with moist western meadows (Bowyer 1984, Bowyer and Kie 2004).

Sexes were strongly segregated during the dry season and aggregated during the wet season (Fig. 31). This initial step demonstrating differences in use of space by the sexes and its pattern on the landscape, which I believe is essential to the definition of sexual segregation, often is neglected in studies of

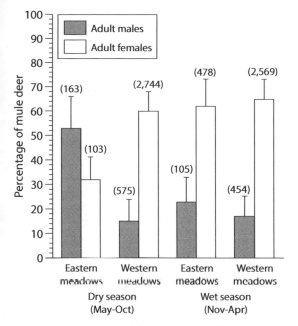

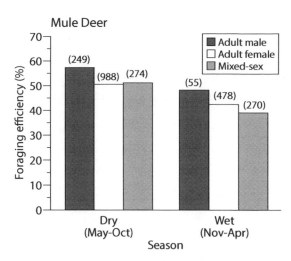

Fig. 31. Percentage of adult male and female mule deer occurring in eastern and western meadow systems on East Mesa, Cuyamaca Mountains, California, during the dry (sexual segregation) and wet (sexual aggregation) seasons. Note that males predominate on eastern meadows and females on western meadows during the dry season. Sample sizes (number of deer) are in parentheses and error bars are +1 SD. (From Bowyer and Kie 2004)

Fig. 32. Foraging efficiency (percentage of active individuals foraging) for mule deer by type of social group and season, corrected for group size with an analysis of covariance, on East Mesa, Cuyamaca Mountains, California. Sexual segregation occurred in the dry season and sexual aggregation in the wet season. Sample sizes (number of groups) are in parentheses. (From Bowyer and Kie 2004)

social segregation. Without such evidence, social segregation cannot claim to be the ultimate explanation for all types of sexual segregation.

Differences in sexual segregation of mule deer also were paralleled by significant increases in mean group size and significant changes in types of social groups, even after lowering alpha to 0.01 to compensate for making multiple observations of deer (Bowyer and Kie 2004). Mixed-sex groups were larger than either male-only or female-only groups during dry and wet seasons. This pattern was especially evident for an increasing percentage of mixed-sex groups from dry to wet seasons (i.e., from parturition to rut) (Bowyer and Kie 2004).

Foraging efficiency (percentage of active deer foraging) differed significantly between seasons and for all comparisons of social groups, except for the appraisal of male-only with female-only groups

during the dry season. Moreover, foraging efficiency of the sexes was most divergent during the wet season, when mule deer formed the most mixed-sex groups and largest social aggregations (Fig. 32). Adult male and adult female deer, however, increased their foraging efficiency by joining mixed-sex groups during both seasons (Fig. 32).

Foraging efficiency was coupled with group size; larger groups foraged significantly more efficiently than smaller ones. When corrected for group size of deer with a covariance analysis, predicted values for foraging efficiency no longer differed between seasons or types of social groups (Fig. 32). Hence, group size was more influential than either sex or season in determining foraging activities of mule deer, an outcome sufficient to reject the hypothesis of activity patterns causing sexual segregation.

Reduced foraging efficiency by males was not likely a result of males acquiring a higher-quality diet during the wet season than during the dry season (Fig. 32). Rutting activities by male ungulates often entail overall increases in activities associated with courtship and aggression. Miquelle (1990) noted that male moose ceased foraging for about an 18-day period at the peak of rut, while engaging in strenuous rutting behaviors. In addition, Barboza et al. (2004) reported that male reindeer lost substantial body mass because of rutting activities. During rut, male Roosevelt elk spent only 25% of their time feeding, whereas adult females fed in 64% of observations (Bowyer 1981). Such rut-related behaviors in male ungulates are commonplace (Apollonio et al. 1989, Komers et al. 1994). Indeed, the greatest disparities in foraging activities between the sexes occur during rut, often in mixed-sex groups, yet those groups are typically the largest of social groups. If differences in activity patterns were the cause of segregation, why is there no effect during the mating season? The activity-pattern hypothesis offers a tenuous explanation for sexual segregation but clearly provides no clarification as to why the sexes aggregate in large mixed-sex groups during rut. Obviously, the costs from any differences in foraging activities between sexes were outweighed by benefits from associating in large, mixed-sex groups. I cannot reconcile hypotheses that postulate that differences in activity between sexes should lead to sexual segregation with observations that ungulates tend to be most gregarious at a time when activities, especially foraging efficiency, of the sexes differ most (e.g., during rut).

At best, hypotheses concerning activity patterns predict short-term avoidance of one sex by the other and do not account for the sexes using the same space (or resources) at different times. Yet, the phenomenon of sexual segregation is characterized by sexes of dimorphic ungulates partitioning use of such space and resources (Bowyer 1984, Miquelle et al. 1992, Bleich et al. 1997, Kie and Bowyer 1999, Mysterud 2000). Moreover, the sexes of some ungulates separate widely, sometimes into distinct mountain ranges—a process that could hardly be caused by differences in activity patterns (Bleich et al. 1997). Indeed, Barboza and Bowyer (2001) argued that variation in activity patterns is an unlikely cause of spatial segregation of sexes, and is more likely a consequence of that behavior.

Criticisms of the Activity-Pattern Hypothesis

In theory, changes in degree of sexual segregation can occur without invoking activity patterns. Differences in body and rumen size between sexes (Chapter 4) can generate changes in preferences for forages or habitats. Such differences are sufficient to produce patterns of sexual segregation on the landscape (Bowyer et al. 2002b; see Table 4), although the gastrocentric hypothesis is necessary to explain periods of segregation and aggregation (Barboza and Bowyer 2000). This outcome also is supported by experimental research on white-tailed deer (Kie and Bowyer 1999). Altering the density of deer via lowering the number of coyotes resulted in marked changes in degree of sexual segregation, with less segregation on the high-density area (see Chapter 4 and Figs. 27, 28), a result I will explain more fully in Chapter 7. One critical point was that this change in the degree of sexual segregation was accompanied by dietary and habitat changes but not by adjustments in types or sizes of social groups (Kie and Bowyer 1999). Importantly, no aspects of social-factors hypotheses were necessary to explain those outcomes, and results run contrary to most social-factors predictions for sexual segregation. Moreover, critical tests of the social-factors hypotheses are difficult, if not impossible, to obtain. Even research on whether the sexes joining or leaving mixed-sex groups affects group activities is equivocal. For example, numerous publications have failed to support predictions from the activity-pattern hypothesis for an array of ungulate species (Mooring et al. 2003, 2005; Bowyer and Kie 2004; Mooring and Rominger 2004; Michelena et al. 2006; and others), as well as for macropods and elephants (MacFarlane 2006, Shannon et al. 2008).

What is the cause of this discrepancy among studies? Neuhaus et al. (2005) acknowledge that social and habitat (ecological) segregation are different independent phenomena. I concur. Those authors also suggest that there have been fundamental problems in naming hypotheses that can confound their interpretation; another point of agreement. Plainly, none of the social-factors hypotheses can explain why the sexes use different spaces during discrete periods of time, the observation that initially led to hypotheses concerning sexual segregation. For example, the activity-patterns hypothesis cannot effectively sort the differences in activity between being a simple correlate of sexual dimorphism or a cause of sexual segregation. The problem herein is that social and ecological sexual segregation are not just independent phenomena, they relate to completely different hypotheses, which is a point in desperate need of clarification.

The Social-Constraints Hypothesis

Hypotheses concerning social factors do not offer an adequate explanation for why the sexes of sexually dimorphic ungulates spatially separate into distinct areas during some seasons and then aggregate in other seasons. What, then, do they explain? I do not perceive major differences between hypotheses forwarded to explain social segregation and those addressing the evolution of sociality (Alexander 1974, Hirth 1977, Molvar and Bowyer 1994, Kie 1999, Bowyer et al. 2001a, Krause and Ruxton 2002). Superficially, these ideas may seem to be opposite sides of the same coin, but they are not. Studies of group living examine the benefits of group formation relative to its costs (Alexander 1974, Krause and Ruxton 2002). In particular, postulates for sexual segregation related to activity patterns and male aggression (Conradt 1998a, Ruckstuhl 1998, Weckerly 2020) are actually hypotheses related to the costs of living in groups. I made this observation nearly 2 decades ago (Bowyer 2004, Bowyer and Kie 2004), but my contention remains largely unaddressed. Indeed, most of the criticisms I have leveled, especially against the activity-patterns hy-

pothesis, have yet to be adequately considered by those proposing social factors as the ultimate cause of sexual segregation.

Although difficult to test experimentally, the logic underpinning the activity-patterns hypothesis appears sound. Sexes with differences in body size, digestive organs, and physiology could reasonably be expected to exhibit differences in forage intake and processing. Such differences in activity by the sexes could lead to costs from joining mixed-sex groups and ultimately to the fission of those groups and resulting fusion into same-sex groups. This process would only result in short-term spatial differences between sexes, and it certainly cannot explain why particular areas are used during sexual segregation. Nonetheless, differences in activity patterns represent a clear cost to living in mixed sex groups. Consequently, I propose that various social-factors hypotheses be recast into the "social-constraints hypothesis." Activities of the sexes then could be evaluated relative to other costs and benefits of sociality (sensu Alexander 1974, Krause and Ruxton 2002). Recognizing the importance of differences in activity patterns between the sexes (Conradt 1998a, Ruckstuhl 1998) and of male aggression (Weckerly 2020) were major advances; those ideas are not currently incorporated into hypotheses for the evolution of sociality (Alexander 1974, Krause and Ruxton 2002), where I believe they make major contributions.

Labeling differences in activity a type of sexual segregation (e.g., social segregation) has been a major source of confusion, and has impeded the development of a common definition for sexual segregation. Numerous papers have attempted to attribute the variation in degree of sexual segregation to lists of potential hypotheses. These laundry lists, however, seldom led to critical tests of hypothesis (Bowyer 2004), and mixing hypotheses for separate phenomena resulted in misperceptions of the processes involved. I believe recasting the activity-pattern hypothesis as the social-constraints hypothesis more clearly represents the biological process under consideration, and it should lead to a clearer path for investigations of group living and sexual segregation.

SUMMARY

This chapter evaluates the primary hypotheses for social factors causing sexual segregation. The shortcomings of each hypothesis are identified with full descriptions of why they are insufficient. Moreover, some of these hypotheses are more closely related to the evolution of group living than sexual segregation, and a new name is provided that more closely parallels their evolutionary meaning.

1. Social-factors hypotheses, which involve affinities of males for one another, males avoiding other males, or females avoiding males as explanations for sexual segregation, lack substantial quantitative support and fail to explain patterns of sexual segregation and aggregation, or why mixed-sex groups should form during both periods.

2. The activity-budget hypothesis for sexual segregation has a number of difficulties, including providing a mechanism for mixed-sex groups to fragment but not for them to merge, and not explaining why mixed-sex groups are typically larger than single-sex groups.

3. Although a logical conclusion, little evidence exists that differences in activity budgets cause fission of mixed-sex groups. This hypothesis most often evaluates activity patterns, because actual time budgets require sampling throughout a diel cycle, which can be challenging.

4. A case study of activity patterns in mule deer demonstrated that group size was more influential than group type (e.g., mixed sex or single sex) in explaining foraging efficiency. These are results that run counter to findings necessary to support the activity-budget hypothesis.

5. An in-depth study of white-tailed deer reported that degree of sexual segregation was unrelated to group size or composition (type of social group), which resulted in rejecting hypotheses related to social factors.

6. Hypotheses related to social factors do not test reasons for why sexual segregation occurs but instead reflect costs of group living; they should be added to hypotheses that explain the costs and benefits of sociality. Consequently, hypotheses related to social factors, and in particular, the activity-budget hypothesis, should be renamed the "social-constraints hypothesis" to better reflect the actual mechanisms, and the social-constraints hypothesis should not be offered as an overall explanation for or definition of sexual segregation.

7 | Ecological Aspects of Sexual Segregation

How are various ecological aspects of the life history of ungulates related to sexual segregation? Ecological aspects of sexual segregation are not divorced from the behavior of ungulates. Degree of sociality can have profound effects on predation risk (Chapter 2), and the behaviors of the sexes as they move to and from particular habitats or areas, and how those patterns vary seasonally (Chapter 3), are critical components in understanding this process. The essential question is not how they accomplish those movements but why the sexes should partition space. Thus, viable hypotheses for sexual segregation should attempt to explain why advantages in reproductive success are conferred to individuals that engage in such movements and behaviors.

The first step in this process is to document that the sexes of sexually dimorphic ungulates make differential use of distinct spaces, including quantifying distances between the sexes and their spatial patterns on the landscape (Chapter 3). When this separation is not pronounced among sexually dimorphic ruminants, examining sampling scale (Chapter 4) or whether partitioning of other resources (this chapter) is occurring may be necessary. The second step is to determine whether sexual segregation differs between seasons—in temperate and arctic ecosystems, segregation of the sexes is most pronounced around the time of parturition and into summer, and sexual aggregation occurs predominantly during the rut (Bowyer 2004; Chapter 3). There are some exceptions, including segregation continuing into winter, aggregations in spring around clumped resources (Bowyer 1984), and grouping of individuals into yards under severe winter conditions (Telfer 1967, 1978; Hodgman and Bowyer 1986). Nevertheless, the changing pattern of segregation and aggregation on the landscape must be present to know whether this phenomenon is occurring. Otherwise, demographic differences that occur between the sexes might be misinterpreted as sexual segregation (Bowyer 2004). For example, polygynous ungulates often display sex ratios that are biased strongly toward females, likely a result of differential mortality of males resulting from reducing feeding during the mating season and engaging in strenuous rutting activities (Bowyer 1981, Miquelle 1990, Bowyer 1991, Berger and Gompper 1999), or a result of preferential harvest of males by hunters (McCullough 1984). Accordingly, noting that males make up a smaller proportion of the population or occur at a lower density than females is not adequate evidence for sexual segregation.

Competition Hypothesis

Clutton-Brock et al. (1987) proposed that intersexual competition was a cause of sexual segregation in red deer, with females competitively excluding males (sometimes termed the *scramble-competition hypothesis* or *indirect-competition hypothesis*). The mechanism underpinning this premise was that grazing pressure by females reduced the availability of preferred forage, and males avoided such areas because of their low tolerance for low standing-crops of grass. Main and Coblentz (1996) likewise concluded that feeding sites of female mule deer had lower biomass of forbs than sites used by males. Further, Illius and Gordon (1987) reported that males were less competitive feeders than females because of allometric differences in the size of bites taken by the sexes. Females, however, typically occur at higher densities than males, and greater removal of forage on female ranges than on those occupied by males would be expected on that basis alone. One major shortcoming of such studies is that demonstrating competition typically requires an experimental approach wherein the density of competitors, or attributes related to their competitive ability, are manipulated (Stewart et al. 2002).

The theoretical arguments for mechanisms of forage processing and intake rates affecting sexual segregation originate from Illius and Gordon (1987)—these were discussed in detail in Chapter 2. Nonetheless, those arguments help form the basis for assertions that sexual segregation is caused by females competitively excluding males, and they require some additional discussion before the role of population density can be further addressed.

Based on captive feeding experiments with Nubian ibex (*Capra nubiana*) Gross et al. (1995) forwarded the hypothesis that mouth architecture, digestive morphology, and forage requirements of the sexes played a major role in sexual segregation. Main et al. (1996), however, questioned whether results presented for ibex were sufficiently broad to provide a general explanation for sexual segregation in ungulates. Gross (1998) responded by outlining

limitations to the sexual dimorphism–body size hypothesis of Main et al. (1996), emphasizing that mechanisms of forage processing and intake rates relating to energetic needs of the sexes were the cause of sexual segregation. In turn, Main (1998) noted that teeth morphology did not provide a strong explanation for sexual segregation and suggested that arguments concerning mouth architecture of grazers differing from those of browsers were moot because both grazers and browsers sexually segregated. As Main (1998) remarked in citing Darwin (1871), when evolution involves both behavior and structure, behavior should precede changes in structure. Indeed, rather than being the cause of sexual segregation in ungulates, differences in mouth morphology likely stem from differences in gastrointestinal morphology and physiology that lead to spatial separation of the sexes into distinct niches (Barboza and Bowyer 2001, Pardi and DeSantis 2021).

Pérez-Barbería and Gordon (2000) examined the mouth morphology of an array of artiodactyls, dominated by species of bovids. Those authors concluded that a difference in body size, rather than mouth architecture, between sexes was the primary factor determining niche differences related to diet, and thereby sexual segregation. Pérez-Barbería and Gordon (2000) correctly noted that avoidance of competition between the sexes was not a main force in the evolution of sexual dimorphism in ungulates (Chapter 2). This conclusion, however, does not mean that partitioning of diet is not involved in sexual segregation, a point I will address later in this chapter.

The lack of differences in mouth architecture between male and female ungulates (see Chapter 2 and Fig. 7) does not completely rule out the possibility of females competitively excluding males via differences in foraging behavior. Spaeth et al. (2004) examined different characteristics of Barclay willow (*Salix barclayi*) twigs on foraging behavior of captive male and female Alaskan moose on the Kenai Peninsula during winter to test experimentally whether females might competitively exclude males, thereby causing sexual segregation. Sexual segregation can be pronounced in Alaskan moose during winter (Miquelle et al. 1992, Bowyer et al. 2001b). Spaeth et al. (2004) reasoned that if females were to exclude males from preferred forages, they would do so by feeding more efficiently than males. Thus, Spaeth et al. (2004) tested whether previous browsing by adult females reduced the foraging efficiency (i.e., intake rate) of adult males, a necessary outcome for competitive exclusion to occur.

Spaeth et al. (2004) obtained data on weight, physical condition (ultrasound rump fat), and size of the incisor arcade for 3 adult male and 3 adult female moose. All females were pregnant (based on ultrasonography) during the course of the experiment, which took place prior to the last one third of gestation; one female, however, did not give birth. Likewise, detailed information was collected on willow twigs, including the length and weight of current annual growth, twig diameter at the bud-scale scar, and data on the density and arrangement of twigs on willow plants. The size of willow twigs affected the absolute and relative sizes of bites taken by feeding moose (Spaeth et al. 2004). Moose took disproportionally larger bites when feeding on smaller twigs than larger ones. Hence, moose consumed relatively more second-year growth of willows when the available first-year growth was limited. Moose likely selected for current annual growth of willows because first-year growth of twigs had a higher nitrogen content than second-year growth (Spaeth et al. 2002). Those patterns of feeding, however, did not differ significantly between the sexes of moose, indicating that differences in mouth architecture resulting in intersexual competition was not a likely determinant of sexual segregation. Weckerly (1993), du Toit (1995), Pérez-Barbería and Gordon (1999), and Spaeth et al. (2001) reached similar conclusions.

The next component of the experiment by Spaeth et al. (2004) was to examine effects of browsing by females on subsequent use of willows by male and female moose. Browsing was evaluated at 2 levels: moderate (22%–50% of current annual growth removed by female moose) and high (51%–78% re-

moved by females) prior to the feeding experiment. Body mass, sex of moose, and density of willow twigs predicted the rate of forage intake (g/sec), with females attaining higher rates of forage procurement than males at moderate and high levels of browsing intensity. Notwithstanding that result, browsing by females did not differentially affect subsequent browsing by males (Fig. 33). Females had a greater

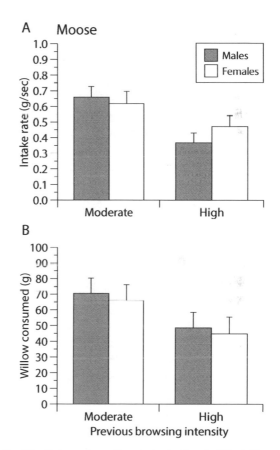

Fig. 33.   Rebrowsing willow twigs by 3 male (18 trials) and 3 female (18 trials) Alaskan moose during winter on the Kenai Peninsula, Alaska, illustrating intake rate (A) and mass of willow consumed (B) at moderate (<51%) and high (>50%) levels of previous browsing intensity (mean + 1 SD). The identity of moose, statistically controlled with analysis of covariance for mass, physical condition, and other idiosyncrasies in foraging behavior, revealed significant differences between sexes in intake rate ($P = 0.0034$) but not for mass of willow consumed ($P = 0.71$). This outcome indicates that, although females foraged more intensively than males, they would have difficulty excluding males via competition for a preferred forage. (From Spaeth et al. 2004)

effect on bite rate of other females than on that of males, especially at high levels of previous browsing intensity. This experiment offers persuasive evidence that exclusion of males by females via foraging behavior is not a compelling explanation for sexual segregation of browsing ruminants.

Other methods exist for assessing competition in ungulates. Indeed, 2 similar publications by members of the research group who originally proposed the scramble-competition hypothesis (Clutton-Brock et al. 1987) examined changes in population density of red deer and concluded that this hypothesis was an unlikely explanation for sexual segregation (Conradt et al. 1999, 2001)—increasing density, which should have intensified competitive interactions, did not lead to increases in degree of sexual segregation. Indeed, du Toit (2005) concluded that there was little overall support for the indirect-competition hypothesis as a cause of sexual segregation.

Moose that were held at low population density by predation, and experienced a high availability of forage as a result, led Miquelle et al. (1992) to conclude that competition was not the primary factor responsible for sexual segregation in moose. Du Toit (1995) reached similar conclusions for kudu (*Tragelaphus strepsiceros*). Kie and Bowyer (1999) conducted a manipulative experiment wherein population density of white-tailed deer in South Texas was increased in an enclosed area by reducing the number of coyotes, while not manipulating coyote populations in surrounding areas (Chapter 4). Coyotes can be efficient predators, especially of neonatal white-tailed and mule deer (Andelt et al. 1987, Bowyer 1987, Lingle 2000, Turner et al. 2011). If competition explained sexual segregation, then the degree of spatial separation should increase at high densities of deer, yet the reverse occurred, with greater sexual segregation at moderate densities; degree of sexual segregation was highest around the time of parturition at both densities of deer (see Fig. 28). Further, greater partitioning of diet at high density was accompanied by a decline in the degree of sexual segregation, which was greater than at moderate

density when sexual segregation was more pronounced (see Fig. 29). This pattern of greater partitioning of resources at higher density has been noted previously (McCullough 1979, Conradt et al. 1999). Moreover, Schroeder et al. (2010) documented a similar pattern for bighorn sheep—greater partitioning of diets between the sexes occurred when the sexes were concentrated on limited winter range by heavy snow in contrast to nearby bighorn sheep at lower elevations that experienced fewer constraints from snow (see Fig. 30). Thus, those authors who examined resources partitioning between the sexes (Weckerly 1993, Jenks et al. 1994) were far closer to understanding mechanisms explaining sexual segregation in ungulates than those who promoted intersexual competition as its cause.

Before discussing underpinnings of resource partitioning further, I need to emphasize why competition is a related but separate process. Competition, whether between sexes or species, is affected by several basic components. Equations for competition incorporate parameters for density-dependent population growth, including the intrinsic rate of increase (r), the ecological carrying capacity (K), and the population size (N) for each competitor (Lotka 1925, Volterra 1926). The Lotka-Volterra equations also incorporate a competition coefficient for each species (α, β). Although oversimplified and not especially predictive, these equations are of substantial heuristic value in understanding the components involved in the competitive process. The key feature I wish to note is that competition increases with increasing population size or density. Consequently, competition cannot be the cause of sexual segregation because the opposite result has been obtained experimentally—increasing population density reduces sexual segregation.

In competition between species, resource partitioning often is viewed as an effect of past competitive interactions (i.e., the ghost of competition past) (Connell 1980), with competitors now occupying more central areas of their respective realized niche spaces. Variation in niche space between species is necessary to avoid competitive exclusion (and thereby extirpation) of one species by the other (Gause 1934, Hardin 1960). Although the sexes clearly have different niche dimensions (see "Niche-Partitioning Hypothesis" later in this chapter), little reason exists for one sex to extirpate the other. The impetus is for coexistence. There is a selective benefit for individual males and females to partition their respective niches.

One final potential for indirect competition between the sexes involves vertical segregation in browsing by giraffe (du Toit 2005). Other ungulates have the opportunity to spatially segregate in a horizontal dimension, whereas giraffe in mixed-sex groups engage in a vertical stratification based on sexual size dimorphism and quality of available browse (du Toit 1990). Browsing intensity by giraffe was sampled on *Acacia nigrescens* trees in South Africa. Leaf biomass available to giraffe increased with height above the ground, and giraffe preferentially browsed shoots that yielded high biomass of leaves (du Toit 1990). Accordingly, large male giraffe derived an advantage from feeding above 2.5 m, which was the height at which small females typically foraged. Females in some environments may select for high nitrogen concentrations and against tannins and fiber in forage; male giraffe are more tolerant of plant secondary compounds such as tannins (Mramba et al. 2017).

Du Toit (2005) extended the idea of vertical stratification in feeding to a number of large herbivores, but evidence that this is a widespread cause of sexual segregation in large herbivores is not compelling—some of those species sexually segregate in a horizontal dimension. Nonetheless, his observations likely offer an example of vertical sexual segregation in giraffe. This result, however, does not rule out niche partitioning between the sexes of giraffe. No experimental evidence indicates this outcome results in competitive interactions that affect sexual segregation.

Main (2008) tried unsuccessfully to resurrect the scramble-competition hypothesis. That attempt disregarded a substantial body of evidence, including experimental studies, and provided no new substantiation to support the contention that scramble competition resulted in sexual segregation. The weight of evidence overwhelmingly indicates that scramble competition is not a viable explanation for sexual segregation. Further investigation of this hypothesis is not a fruitful endeavor.

Gastrocentric Hypothesis

Separation in diet between the sexes of ungulates exists, in part, because of sexual dimorphism in size and differences in the digestive morphology and physiology of the sexes. The gastrocentric model (Barboza and Bowyer 2000, 2001) proposes that male ruminants consume low-quality but more abundant forages because their large rumen-reticulum increases retention time, which allows for greater digestion of those forages than for nonpregnant females (see Chapter 2 and Fig. 8). Smaller females, however, are better at postruminal digestion of food; forage intake increases with energy and protein requirements for reproduction (Barboza and Bowyer 2000). Females undergo a remodeling of their digestive tract to support the increased demands of late gestation and lactation (Barboza and Bowyer 2000, Zimmerman et al. 2006; see Fig. 8). Thus, differences in feeding activity between the sexes are a result of those metabolic demands rather than its cause (Barboza and Bowyer 2001).

Moreover, males cannot switch quickly between diets of differing quality because those changes would upset ruminal fermentation, risk excess production of gases, and promote bloat, malabsorption, and scouring (Van Soest 1994, Gordon and Illius 1996, Barboza et al. 2009). This hypothesis offers an explanation for seasonal changes in diet selection and foraging behavior of the sexes, and it helps explain why dominant males do not displace subordinate females from higher-quality ranges (Barboza and Bowyer 2000; Chapter 2). Males can obtain sufficient nutrition from lower-quality and more abundant foods than those required by reproducing females (Barboza and Bowyer 2000), an outcome that provides a credible explanation for sexual segregation.

Niche-Partitioning Hypothesis

Bowyer and Kie (2004) proposed a synthetic approach for understanding sexual segregation in ungulates that incorporates niche theory (MacArthur 1968) and involves existing concepts concerning spatial, dietary, and habitat separation by the sexes (Mysterud 2000). Coexistence of vertebrate species with similar niche requirements necessitates that resources be partitioned. If there is strong overlap on one niche axis (e.g., food), then there must be avoidance on another axis (e.g., space) for competitive exclusion to be avoided (Schoener 1974). Indeed, 2 species cannot occupy the same niche space at the same time (Gause 1934) without one being competitively excluded (Hardin 1960).

The sexes of dimorphic ruminants, which have somewhat different dietary requirements based on body size, gut morphology, and physiology, also may partition resources and space (see Table 5). Treating spatial, dietary, and habitat differences between sexes in a niche context (see Table 5) further develops the ideas of Mysterud (2000). This approach allows integration of existing concepts to provide a unique method for understanding sexual segregation. Among the sexes of ungulates, overlap on one niche axis often is accompanied by avoidance on another, which is a key element in sexual segregation (Kie and Bowyer 1999, Stewart et al. 2002, Schroeder et al. 2010). During sexual segregation, males and females have markedly different niche requirements that are critical components of their ecology. Management of habitat needs to consider such differences between the sexes, and numerous authors have suggested the sexes of ungulates should be considered and managed as if they were separate species (Kie and Bowyer 1999, Stewart et al. 2003b, Bowyer 2004, Schroeder 2010, Whiting et al. 2010, Stewart et al. 2011; Chapter 9).

Considering basic niche requirements (i.e., space, diet, and habitat) separately has led to uncertainty about the causes of sexual segregation; many authors investigated only a single aspect of this complex phenomenon. Moreover, this lack of an integrated approach has led to the conundrum that, in some studies, sexes were identified as segregating, or failing to do so, based on the examination of only one of several potential niche axes. Studying only a single facet (i.e., niche axis) of sexual segregation or treating interrelated and interacting axes as independent variables offers an incomplete overview of this process and its underlying mechanisms. This lack of an integrative approach has led to considerable confusion over which species sexually segregate and why they do so.

Bowyer et al. (1996) observed that during sexual segregation male and female black-tailed deer exhibited little divergence in use of habitats. Others also have noted a lack of difference in habitat selection or small differences in diet between sexes of ruminants when they were segregated (Beier 1987, Cransac et al. 1998, Conradt 1999, Fulbright et al. 2001). Nonetheless, I urge a more integrative approach that synthesizes the multiple factors related to sexual segregation. Testing the niche-partitioning hypothesis offers such an avenue of investigation.

A number of reasons exist for not detecting sexual segregation in sexually dimorphic ruminants that engage in this behavior. For example, sampling may occur at the wrong scale to identify this process (Bowyer et al. 1996, Kie and Bowyer 1999, Bowyer and Kie 2006). Nearly any answer as to whether the sexes segregate can be had by varying sampling scale (see Fig. 26). For white-tailed deer that partition space on a relatively fine scale (1–4 ha), segregation can be easily overlooked (Kie and Bowyer 1999). The niche-partitioning hypothesis offers another reason for sexual segregation to go unnoticed—space, habitat, and dietary differences between the sexes all need to be considered before concluding that sexual segregation does not occur. Niche axes other than space, habitat, and diet are also possible, although not likely as important across the diverse species of ungulates that engage in sexual segregation. For example, temperature and availability of free water are probably more important axes in some species and environments than in others, although such variables are clearly components of habitat. The basic categories of space, habitat, and diet must be considered along with these minor niche axes, which are subsets of the overall habitat axis. Sexual segregation has been identified among all sexually dimorphic ruminates for which it has been studied in detail (Chapter 2); thus, attributes unique to a few species in particular environments do not offer an explanation for this ubiquitous phenomenon (Bowyer 2004).

Our understanding of sexual segregation has been hindered by simply describing the phenomenon rather than testing why it occurs (Bleich et al. 1997). Moreover, Bowyer (2004) cautioned that numerous studies of sexual segregation likely were not designed specifically to test for spatial separation of the sexes. Indeed, some publications may have used data collected for other purposes to examine sexual segregation, often failing to obtain critical tests of hypotheses or assuming that finding one correlate of sexual segregation excludes other potential causes. In addition, many studies did not consider seasonal variation in degree of sexual segregation as part of this phenomenon (Bowyer 2004). Little has changed with respect to those practices over the past 3 decades. Indeed, some publications on sexual segregation likely involved HARKing—hypothesizing after results are known—a procedure contrary to standards of the scientific method (Kerr 1998). One additional factor has potential to affect interpretation of differences in niche requirements between the sexes: predation risk.

Predation-Risk Hypothesis

Effects of predation risk are pervasive among ungulates, with marked differences occurring between the sexes. Such effects can occur even if predators are not currently present—ungulates have a long evolutionary history of interactions with large carnivores that prey upon them and may react as if predators were present even when they are not (Berger and Gompper 1999, Berger et al. 2001c, Bowyer 2004). Pronounced sexual size dimorphism leads to differences in vulnerability to predation between adult males and females (Chapter 2). In addition, such differences in size may make smaller females more vulnerable than larger males to a wider range of predators. Moreover, females and their attending young are highly susceptible to predation until neonates can begin escaping from predators (Bergerud and Elliot 1986). Those factors promote differences in behaviors, diet, and habitat selection between the sexes of ungulates, which can lead to sexual segregation.

Among ungulates, sex ratios of adults are skewed toward females, largely because of increased mortality of males (Leader-Williams 1988, Bowyer 1991, Kie 1999). Polygynous males tend to engage in more risky behaviors than females (Jones 2014) and participate in rutting activities that often lead to exhaustion, malnutrition, and sometimes death (Bowyer 1981, Apollonio et al. 1989, Miquelle 1990, Barboza et al. 2004). Ungulates in poor condition or with debilitating disorders often are selected as prey by predators (Peterson 1977, Mech et al. 1995). Females and young are more vulnerable to predators than males, but females engage in behaviors that lessen the probability of being selected as prey. In open-land ungulates, females may form large groups (Bowyer et al. 2007, Fortin et al. 2009) or select rugged and steep terrain, where the likelihood of successful pursuit by predators is lessened (Bleich et al. 1997, Rachlow and Bowyer 1998, Robinson et al. 2020). Females and young also may remain closer to escape terrain than males (Berger 1991, Schroeder et al. 2010). Females similarly seek areas with fewer predators during and following parturition (Bergerud et al. 1984, Bergerud and Page 1987, Bleich et al. 1997, Barten et al. 2001; see Figs. 11, 17). Likewise, females may actively defend themselves and young from predators (White et al. 2001, Grovenburg et al. 2009), although their nutritional condition may

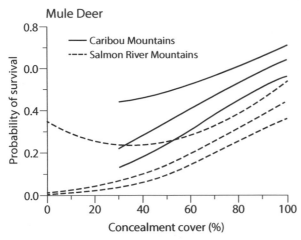

Fig. 34. Probability of survival (±SE) from birth to 5 months of age for young mule deer as a function of percent concealment cover at 0–50 cm in height at neonatal bed sites measured at birth and at 4 to 6 weeks postparturition in the Caribou Mountains (*n* = 43) and the Salmon River Mountains (*n* = 51), Idaho. Predictions are based on the best model that included stage-specific survival, study area, and percent concealment cover. (From Shallow et al. 2015)

determine how willing they are to defend young from predators (Smith 1987). Females and especially their young may engage in cryptic behaviors to avoid predators by using concealment cover (Fichter 1974, Kitchen 1974), which enhances survival of neonates (Jacques et al. 2015, Shallow et al. 2015; Fig. 34). Female African elephants and young occur in denser cover than males, which inhabit more open areas, likely a response to predation risk (Kioko et al. 2020).

Some females may attempt to negate hunting tactics of predators by making extensive movements immediately prior to parturition (Bowyer 1999b). Parturition in many ungulates is highly synchronized (Bowyer 1991, Bowyer et al. 1998b, Rachlow and Bowyer 1998, Long et al. 2009a). Synchronized births are thought to swamp predators with more young than can be killed, and thereby serve as an antipredator strategy (Estes 1976; Estes and Estes 1979; Rutburg 1984, 1987). Evidence supporting this hypothesis is not widespread. Bowyer (1998b) failed to

find antipredator benefits in highly synchronized births of moose; at north latitudes, climatic patterns may be more important than predators in synchronizing births. Moreover, desert-dwelling bighorn sheep have births distributed across a protracted period (Bleich et al. 1997). Synchronization of births may not be as an effective antipredation mechanism as previously thought. As an overall pattern, however, females ameliorate their susceptibility to predation risk via differing behaviors and selection of specific environments that reduce the risk of predation, which can lead to sexual segregation.

Multiple Causations and Tradeoffs

Neuhaus et al. (2005) and Main (2008) proposed that multiple causes were necessary to explain sexual segregation, including a long list of both social and ecological factors. All of these potential causes would not be expected to operate for all species across diverse environmental circumstances, and they fail to provide a clear path to testing among varied hypotheses (Stewart et al. 2011). If an extensive list of hypotheses was necessary to explain sexual segregation, multiple examples should exist among dimorphic ruminants where sexual segregation did not occur, a pattern contrary to numerous observations. Moreover, the best that can be hope for under such an approach is to attribute the relative amounts of sexual segregation to various hypotheses—something that will not provide a critical test of why sexual segregation occurs (Stewart et al. 2011). Moreover, some of these tests, such as those for activity patterns, are related to hypotheses other than sexual segregation (Chapter 6).

I have argued previously (Bowyer 2004) that there may be no more than 2 general hypotheses necessary to explain sexual segregation in ungulates: resources and predation. Resources and predation also are suitable for examining the niche-partitioning model for sexual segregation. Considering how these hypotheses interact is necessary to elucidate patterns of spatial separation of sexes on the landscape (Bowyer

2004). These hypotheses are independent but not mutually exclusive; either resources, predation, or both may affect sexual segregation. Tradeoffs between acquiring essential resources and avoiding predation are well documented for ruminants (Festa-Bianchet 1988, Berger 1991, Molvar and Bowyer 1994, Rachlow and Bowyer 1998, Kie 1999, Barten et al. 2001, Corti and Shackelton 2002); sometimes those tradeoffs may result in females acquiring a lower-quality diet than that of males, as long as it satisfies requirements for reproduction and lactation (White and Luick 1984, Bowyer 2004). Failure to evaluate tradeoffs can cause misinterpretation of results from studies of sexual segregation.

A conceptional model illustrating tradeoffs (or the lack thereof) between predation risk, risk of nutritional deficiency, and diet quality for males and females during sexual segregation is illustrated in Figure 35. Females are assumed to obtain a higher-

quality diet than males (Barboza and Bowyer 2000, Barboza et al. 2018). Tradeoffs between predation risk and nutritional deficiency, however, may result in males obtaining a higher-quality diet than females (Fig. 35). Thus, predictions from the gastrocentric model may be modified by risks of predation and nutritional deficiencies. Females obtaining a lower-quality diet than males is not sufficient evidence to reject the gastrocentric hypothesis; instead, dietary requirements may be interacting with predation and nutritional risks to influence the quality of food for the sexes (Fig. 35). Obviously, other factors can influence the quality of food obtained by ungulates, including insect harassment and anthropogenetic disturbances (Toupin et al. 1996, Cameron et al. 2005).

This conceptual model (Fig. 35) provides a framework for understanding how various aspects of ungulate behavior and ecology are related to sexual

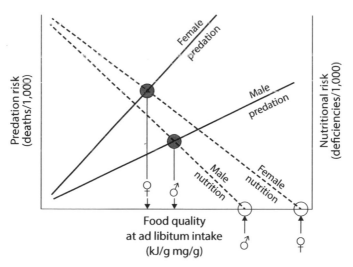

Fig. 35.   A conceptional model for relationships between predation risk and forage quality for male and female ungulates during sexual segregation with the assumptions that females require a higher-quality diet and are more vulnerable than males to predators. Inverse dashed lines indicate no constraints from predation risk on ungulates from obtaining maximum forage quality, whereas the positive solid lines indicate that animals experience reduced quality of forage because of tradeoffs with predation risk. Optimal solutions of minimizing the predation to forage ratio (*open circles on x axis*) identify points where forage quality is maximized and predation risk minimized for males and females. Positive lines for the sexes indicate tradeoffs between forage quality and predation risk wherein constraints are placed on forage quality because of predation risk (*gray circles* show the intersections of lines resulting from tradeoffs for males and females). Upward-pointing arrows on the *x* axis show differences in forage quality obtained by males and females with no tradeoffs, and downward-pointing arrows indicate quality of diets obtained by the sexes when tradeoffs occur. (Courtesy of P. Barboza)

segregation. The model is not meant to be predictive, and the slopes of lines in the model are intended as generalizations. Moreover, there is no reason that the lines need to be linear. In addition, forage quantity could have been modeled instead of diet quality (males require a greater quantity of forage than females). Also, examining how predation and nutritional risks changed with age might be useful (and certainly would change the slopes of the lines). Nonetheless, the conceptual model suffices to produce a take-home message that predation and resources are the primary mechanisms likely to explain sexual segregation in ungulates.

Are there examples where predation risk is minimized? Recall that predators need not be present to engender a response to risk in the environment. Moreover, females may engage in behaviors that reduce risk, such as seeking rugged and steep terrain to avoid predators or occurring on ranges where the number of predators is diminished. This outcome would reduce the steepness of the tradeoff line (Fig. 35) and potentially improve diet quality. Rachlow and Bowyer (1998) reported that parturient Dall's sheep selected small level areas within rugged and steep terrain to give birth, and that those areas also were near high-elevation patches of *Dryas*, a preferred forage. Merrill et al. (2020) noted that North American elk exhibited tradeoffs between forage and predation risk and that those relationships were affected by population density. Gedir et al. (2000) noted that tradeoffs between forage and predation risk could be affected by drought in bighorn sheep. Heffelfinger et al. (2020) reported the absence of a tradeoff between forage and predation risk for mule deer, and Bowyer et al. (1998a) failed to find a tradeoff for neonatal black-tailed deer between forage and risk of predation. Bowyer et al. (1999b) noted that Alaskan moose selected dense patches of willow for birth sites that offered concealment cover for neonates but also provided females with high-quality forage. Nevertheless, those authors also reported that there were random sites with more willows than those selected by maternal moose, in-

dicating that predation risk may still have been involved. Predation risk is pervasive for ungulates, but the absence of a tradeoff between forage and predation risk probably relates to specific environmental conditions. Predation risk has obvious importance for exploring the role of sexual segregation in evolutionary theory, and for the conservation of ungulates.

SUMMARY

This chapter discusses the primary ecological hypotheses for explaining sexual segregation in ungulates, including the scramble-competition hypothesis, the gastrocentric hypothesis, the niche-partitioning hypothesis, and the predation-risk hypothesis. Multiple cautions and tradeoffs between hypotheses are also discussed, as are some reasons why critical tests of sexual segregation are difficult to obtain.

1. Sexual segregation is not detached from the behavior of ungulates. Sociality affects predation risk, and the sexes make seasonal movements that result in spatial separation. The central question, however, is not *how* they accomplish such movements but *why* they should do so.
2. There may be no more than 2 general hypotheses needed to explain sexual segregation: resources and predation.
3. Scramble competition has been proposed as a reason for the sexes to spatially segregate. Nonetheless, even the research group that initially proposed that hypothesis has indicated that it is unlikely. Evidence that differences in mouth morphology between the sexes lead to sexual segregation has been rejected on several occasions, sometimes with experimental evidence. Increasing population size or density should amplify competition between the sexes and thereby the degree of sexual segregation, but the reverse occurs. Competition can no longer be considered a viable hypothesis for sexual segregation.
4. The gastrocentric hypothesis is based on the degree of sexual dimorphism and differences in

digestive morphology and physiology between males and females. Females reconstruct their digestive tract to better accommodate the needs of late gestation and lactation, which results in them consuming a higher-quality diet than males. Males, however, consume low-quality but more abundant forages. Those differences in nutritional requirements can produce sexual segregation.

5. The niche-partitioning hypothesis integrates spatial, dietary, and habitat selection by the sexes to understand sexual segregation. Partitioning of space, diet, or habitat by the sexes can result in spatial segregation of the sexes. If overlap in one niche axis such as diet occurs, then there must be partitioning on another axis such as space. The failure to consider multiple niche axes can lead to misconceptions about whether sexual segregation is occurring. Moreover, the niche requirements of the sexes can be sufficiently different that males and females should be considered and managed as if they were separate species.

6. The predation-risk hypothesis invokes differences between males and females and their young in vulnerability to predators. Females with young often use rugged and steep terrain or form large groups to limit their susceptibility to predation. Females also may select areas with fewer predators for giving birth. Males can use areas with higher predation risk to secure higher-quality or more abundant resources. These processes can lead to spatial segregation of the sexes.

7. There may be tradeoffs between predation risk and forage quality during sexual segregation. Risk may cause females to lower their quality of diet to avoid predators. Males and females should seek areas that lower the predation risk to forage ratio, but such conditions may be difficult to find in nature.

8 | Consequences of Sexual Segregation for Theory and Management

How is sexual segregation integrated into concepts relevant to theory and management of ungulates? Spatial segregation of the sexes is intertwined with theories related to parental investment and population ecology. These topics underpin a broad foundation in evolutionary biology and ecology, but the interrelated roles of parental investment and population dynamics among ungulates have not been fully recognized or widely appreciated. The purpose of this chapter is to amalgamate the concepts of parental investment with population dynamics of ungulates to highlight the importance of sexual segregation in those processes (see Fig. 4).

Paternal Investment and Sexual Selection

Our understanding of sexual selection, mating systems, and parental care expanded rapidly in the later one-half of the past century (Clutton-Brock 1991). Bateman (1948) observed that female fertility was much more limited than that of males; production of larger eggs by females was more energetically expensive than production of smaller sperm by males. Such differences in gamete size (*anisogamy*) and expense were proposed to result in the undiscriminating eagerness of males for copulating and the more discriminating nature of females. Bateman

(1948) demonstrated that males exhibited greater variation in reproductive and mating success, and displayed a stronger relationship between mating success and reproductive success than females (*Bateman's Principles*), which has importance for interpreting sexual selection (Arnold 1994, Jones et al. 2002).

Female mammals show a reduced willingness to reproduce unselectively in comparison with males—a consequence of the greater physiological and energetic expense incurred by females in the production of each successful offspring (Williams 1966). For females, copulation may entail a protracted commitment to a reproductive burden with associated costs and dangers, whereas for males there is far less of an obligation and fewer risks—males exhibit a willingness to mate with as many females as are available, and females are more choosey in selection of mates, especially among mammals (Williams 1966). Such differences in behavior are relevant to understanding relationships among parental care, competition, and sexual selection.

Trivers (1972) coined the term *parental investment* to mean any investment in an individual offspring that increases the chance of survival and reproductive success of that offspring at the cost of investing in future offspring. Thus, the degree of parental investment is modulated by the magnitude of its negative effect on producing future young. Moreover, the relative investment in offspring by males and females influences the amount of competition within the sexes and thereby the intensity of sexual selection (see Fig. 4). Trivers (1972) proposed that parental investment was the essential factor promoting sexual selection. The sex providing the least parental care competed most intensively for mates, resulting in strong sexual selection and providing an evolutionary pathway to sexual dimorphism (Andersson 1994; see Fig. 4).

Parsing the relative effects on sexual selection of Bateman's Principles (Arnold 1994, Jones et al. 2002), parental investment (Trivers 1972), differences in operational sex ratios (Emlen and Oring 1977), and potential reproductive rates of the sexes (Clutton-Brock and Parker 1992) is beyond the scope of this chapter and would require several tomes to address adequately. Indeed, Clutton-Brock (1991) published a book on the evolution of parental care, and Andersson (1994) a text on sexual selection. My interest is in understanding how paternal investment plays a role in sexual selection and population dynamics, and in understanding the relationship of sexual segregation to those processes.

Sexual selection is thought to be the primary cause of sexual dimorphism in mammals (Chapter 2), a proposition consistent with hypotheses concerning the role of male-male competition in promoting differences in body size and weaponry among ungulates (Bro-Jørgensen 2007). This contrasts somewhat with Trivers' (1972) proposition that parental investment was the fundamental factor driving sexual selection. For instance, direct paternal care is lacking in the monomorphic and monogamous Kirk's dik-dik (Komers 1996). Among ungulates, monogamy is not always linked with paternal investment. Indeed, Lukas and Clutton-Brock (2013) contended that parental care in mammals was a consequence rather that a cause of monogamy.

Where degree of polygyny is strongly related to the magnitude of sexual size dimorphism in ungulates, instances can occur when a strong feedback

mechanism exists that further limits the opportunity for paternal care of young (Bowyer et al. 2020a; see Fig. 4). The sexes of dimorphic ruminants have evolved elaborate differences in their digestive systems to meet essential life-history requirements (Barboza and Bowyer 2000, 2001). Hence, males and females of dimorphic ruminants typically are separated spatially from one another for some portion of the year, especially around parturition and lactation (Bowyer 2004). Moreover, predation risk may promote spatial separation of the sexes, with females seeking out steep and rugged terrain for giving birth (Chapter 2). Some species sexually segregate into mountain ranges that are >15 km apart (Bleich et al. 1997) or, for chiru, separated by hundreds of kilometers (Schaller 1998). The time spent sexually segregated varies among species (Bowyer 2004) but is sufficient to prevent extensive paternal investment. Under those circumstances, there are few opportunities for males to recognize their offspring or for the development of paternal care. Because of sexual segregation, selection for male-male combat ostensibly is intensified, augmenting the evolution of sexual size dimorphism and influencing various mating systems in ungulates (Bowyer et al. 2020a; see Fig. 4). Indeed, Székely et al. (2000) proposed that important feedbacks occurred between mating strategies, mating opportunities, and patterns of parental care. Sexual segregation underpins such relationships in dimorphic ungulates.

Population Dynamics

Differences in body size, digestive morphology and physiology, selection of forages, and risk of predation are primary causes of sexual segregation in ungulates (Bowyer 2004). A critical outcome from these sexual differences in use of space and other resources is that females compete more strongly with one another (and young) for forage than they do with males—intrasexual competition among females is intense (see Fig. 4). Moreover, during periods of sexual segregation, females occur at much higher densities than males because of differential mortality between the sexes, which further exacerbates competition among females and between females and young. I have argued that the sexes of ungulates should be treated and managed as if they were separate species (Chapter 7). A primary method of managing ungulates involves harvest (Bowyer et al. 2014, 2020b), which contains counterintuitive components stemming from sexual segregation.

Sexual segregation results in females being more intensively involved in density-dependent processes regulating ungulate populations than males (see Fig. 4). A harvest of males does little to affect population productivity relative to a reduction in females (McCullough 1979, Boyce 1989, Freeman et al. 2014). Some consequences from harvesting male ungulates exist (Mysterud et al. 2002) but are far less pronounced than those of reducing the number of females (Bowyer et al. 2014). Harvesting males to such low numbers that females are not fertilized is uncommon (Schwartz et al. 1992, Laurian et al. 2001, Freeman et al. 2013), largely because of the polygynous mating systems of these sexually dimorphic species (Bowyer et al. 2020a). A particularly heavy harvest of males, however, will reduce their age structure (i.e., result in a population with more younger males); younger males tend to have smaller body size and horn-like structures than older males (Jenks et al. 2002, Monteith et al. 2013a). An undesirable outcome of harvesting only males is that mating by young males is spread over a longer interval than for older males, which ostensibly results in less synchronous birthing (Noyes et al. 1996).

Ungulates exhibit strong density dependence (Chapter 1). A decline in vital rates (e.g., survivorship, reproduction) occurs with increasing population density of females (density of males is largely inconsequential). Such changes occur, in part, because of the slow-paced life-history characteristics of these large herbivores (Chapter 1). Specific demographic attributes that change with the size of the population relative to K (ecological carrying capacity) include survival of young, age at first reproduction, litter size,

weight of neonates, and survival of adults (Albon et al. 1983, Gaillard et al. 2000, Keech et al. 2000, Eberhardt 2002, Bowyer et al. 2014, Monteith et al. 2014b, Gilbert et al. 2020). These changes in density-dependent characteristics of ungulates relative to the size of the population with respect to K have important implications for understanding their population dynamics.

Density-dependent demographics of ungulate populations in which limitation of food results from competition among females produces a decline in their nutritional condition (Monteith et al. 2014b, Oates et al. 2020). This outcome contrasts markedly with species exhibiting strong density independence. Such species tend to be small-bodied, have high reproductive rates, and have short lifespans. The growth curves of density-independent populations tend to be J shaped—increasing populations seldom reach K and, hence, do not experience food limitation before being markedly reduced by density-independent events, such as severe weather, to low levels from which they subsequently recover (Leopold 1933, Bowyer 2020b). The individuals that perish from severe weather constitute Leopold's (1933) "doomed surplus"—animals available to harvest because they would die from other mortality factors anyway if not harvested (i.e., the harvestable surplus). Because of differences in their life-history characteristics (Chapter 1), the population dynamics of species experiencing density independence hold limited value for interpreting the dynamics of ungulate populations.

Some approaches for assessing density dependence in ungulates are inherently flawed. Comparing densities among populations to determine variables such as recruitment or mortality is of questionable utility (Bowyer et al. 1999a, 2014, 2020b). For example, if 5 hypothetical populations of deer occurred in habitats that exhibited substantially different ecological carrying capacities (K), and those 5 populations were at or near K, similar population characteristics such as recruitment or survival would be expected (Table 6). A regression of percent of overwinter mortality against population density, however, would lead to the conclusion that the dynamics of the populations were density independent and being regulated by winter conditions (predicted from a random pattern or a linear fit with a regression slope near zero). Yet, density dependence clearly was regulating those populations, which were all near K; even though no obvious relationship existed between population size and overwinter mortality (Table 6). Notably, habitat quality can confuse comparisons of density dependence among populations of ungulates if K is unknown (Kie et al. 2003). Clearly, population density can be a misleading indicator of habitat quality (Van Horne 1983). The foregoing examples demonstrate how data related to density dependence might be perceived incorrectly and misinterpreted.

Another misconception concerning K is that it is a seasonal phenomenon (Bowyer et al. 2014). Winter, which is a season of forage limitation for temperate and arctic ungulates, has been a focus for research on ungulate populations (Mautz 1978, Bergman et al. 2015). Accumulating evidence, however, demonstrates that nutritional quality of summer and autumn ranges also are important in ungulate population dynamics (Cook et al. 2004, Stewart et al. 2005, Couturier et al. 2009, Monteith et al. 2013b, Loe et al. 2021). Stewart et al. (2005) provided experimental evidence of the importance of summer nutrition on pregnancy rates and nutritional condition of North American elk by manipulating population density. Likewise, Tollefson et al. (2010, 2011) demonstrated the importance of summer nutrition on reproduction, growth, and survival of young mule deer by manipulating summer diets. Nonetheless, nutritional contributions from seasonal ranges are not independent; carryover effects, including those related to provisioning of young, influence the nutritional state of an individual entering the next season (Bårdsen and Tveraa 2012; Monteith et al. 2013b, 2014b; Merems et al. 2020). Consequently, which seasonal range is more important in population regulation is arguable. High-quality forage resulting in enhanced nutritional condition in a particular season may help compensate for the poor forage and condition in another—thus, year-round contributions to K are important in dimorphic ungulates (Kie et al. 2003, Monteith et al. 2013b, Oates et al. 2020).

Many polygynous ungulates are capital breeders (Apollonio et al. 2020), where such carryover effects between seasons would be expected. A capital breeder obtains resources in advance and stores them endogenously until they are needed for activities associated with production of offspring (Barboza et al. 2020). Conversely, an income breeder adjusts food intake concomitantly with breeding, without relying on body reserves (Jönsson 1997). Species employing these foraging tactics lie along a capital-to-income breeder continuum (Williams et al. 2017). For ungulates, reproductive demands of females are high in late gestation, at birth, and during lactation, whereas demands on males are associated mostly with strenuous rutting activities. For income breeders, there would be little expectation of carryover effects between seasons, perhaps intensifying seasonal limitations in quality and abundance of forage, or limited resources for populations at or near K.

The S-shaped growth curve of density-dependent ungulates is related primarily to the nutritional status of females (McCullough 1979, Fowler 1987, Bishop et al. 2009). As population size increases from small to large, per capita availability of food declines for females because of intense intrasexual competition,

Table 6. Carrying capacity (K), population size, number dying, and overwinter mortality in 5 hypothetical populations of deer inhabiting ranges of varying quality (from Kie et al. 2003)

Population	K (number of deer)	Population size (n)	Number dying	Overwinter mortality (%)
A	307	321	30	9.4
B	512	507	50	10.0
C	114	124	12	9.7
D	357	362	36	10.0
E	200	214	22	10.3

which eventually results in negative effects on reproduction and survival. Those density-dependent effects become most pronounced once abundance surpasses the inflection point of the S-shaped curve and begins to asymptote at K (McCullough 1979, 1999; Monteith et al. 2014b). Population irruptions also may result in overshoots of K (Leopold 1943, Klein 1968, Caughley 1970, McCullough 1997, Ricca et al. 2014); K subsequently may be lowered from overgrazing of ranges (Fig. 36). The general concept of what constitutes K is conditioned by the logistic equation (Verhulst 1838), where K is assumed to be a constant parameter. Increasing ungulates populations, however, most likely reduce K as they grow, resulting in a pattern more typical of the dashed line representing a loss of K in Figure 36. Whether or how quickly K recovers from a population irruption and overshoot depends on the resilience of that ecosystem to overgrazing, and the time span over which forages were overgrazed. Nonetheless, a small change in K may precipitate a large change in ungulate numbers (McCullough 1979, Bowyer et al. 2005). Such an outcome results from the nonlinear relation between annual recruitment and population density with respect to K. Hence, a loss of K can disproportionally reduce the overall productivity of the population

(Bowyer et al. 2005); all nonintuitive outcomes related to sexual segregation (see Fig. 4).

Substantial empirical evidence documents the widespread occurrence of strong density dependence among sexually dimorphic ungulates (McCullough 1979, 2001; Sauer and Boyce 1983; Kie and White 1985; Fowler 1987; Boyce 1989; Stewart et al. 2005; Bonenfant et al. 2009; Monteith et al. 2014b). A valuable method for visualizing how density dependence affects population dynamics of ungulates involves graphing the number of recruits (the number of young successfully added to the population in a reproductive effort) against population size (McCullough 1979, Fowler 1987; Fig. 37). This Ricker-like parabola shows that number of young added to the population is low at extremely low numbers because too few adult females exist to produce offspring, and low at high numbers (near K) because females are in poor condition and the few young produced have poor survival (McCullough 1979). Physical condition can be determined by ultrasound measures of rump fat (Monteith et al. 2014b) or from visual inspection of physical characteristics (Berger 2012, Shallow et al. 2015, Smiley et al. 2020). Most young are recruited into the population at intermediate numbers, a point termed maximum

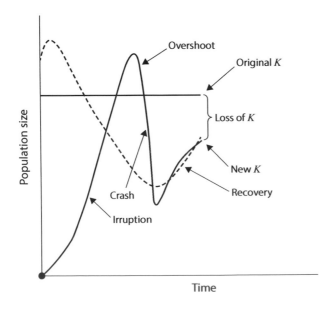

Fig. 36. A hypothetical growth curve for white-tailed deer on the George Reserve, Michigan, illustrating an overshoot of K, a loss of K from overgrazing, a population crash, and recovery to a new equilibrium (*solid line*) (from McCullough 1979). The dashed line illustrates that K is not constant and is reduced with increasing population size. The initial increase in K at low density is caused by herbivore optimization.

Density-Dependent Recruitment

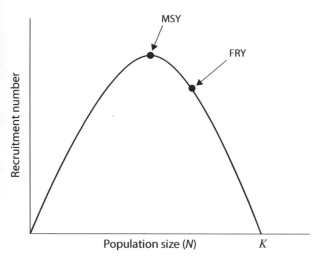

Fig. 37. The parabolic relationship between recruitment number (i.e., the number of young successfully added to the population) and population size for an ungulate population. MSY is maximum sustained yield, which is the maximum harvest (or other mortality) that can be sustained by the population, FRY is a fixed removal yield, and *K* is the number of individuals that the environment can support under equilibrium conditions. The yield curve is derived from the near-linear inverse relationship between recruitment rate (young/adult) and population size (McCullough 1979). (Adapted from Bowyer et al. 2014)

sustained yield (MSY; Fig. 37). MSY is the product of population size and recruitment rate (young/adult), and it lies at the peak of the recruitment parabola (Fig. 37). MSY also indicates the maximum annual harvest (or deaths from other causes) that a population can sustain under a particular set of ecological conditions without causing a decline in numbers (McCullough 1979; Bowyer et al. 2014, 2020b). Density-dependent recruitment is what allows the sustainable harvest of ungulate populations, because recruited young compensate for adults removed by deaths. Consequently, the age structure of the population becomes younger with an increasing harvest that reduces population size relative to *K* (Bowyer et al. 1999a). Note that it is number of adult females relative to *K* and not population density per se that is the critical factor. Sexual segregation underpins such results.

Interpreting outcomes from the recruitment curve (Fig. 37) can be counterintuitive. I have had many discussions with hunters, given presentations to sport-hunting organizations, and attended meetings of game boards in several states. No amount of common sense will allow a clear perspective on how density dependence operates in ungulates. The most common concern and potential misunderstanding is that not harvesting females will always lead to in-

crease productivity of the population. The rationale is typically that 1 female produces 2 young, which in turn reproduce, with their subsequent offspring further adding to the population in an exponential fashion. An often-heard adage is, "Everyone knows that you don't shoot your breeding stock." If a population is at low numbers in relation to *K*, then females will be in good physical condition because of lax competition and an abundant food supply. Most females will be pregnant and litter size (when it is variable) will increase; the size of neonates also will be large and their survivorship high. If the objective is to increase the population, then females should not be harvested and (in the absence of other mortality factors) the population will grow toward MSY (Fig. 37)—in this instance, common sense prevails.

If this same population is near *K*, however, a different set of circumstances occurs (Fig. 37). Females in a population near *K* are in poor physical condition, have low reproductive rates, and produce small young with low survivorship. If the harvest is reduced to compensate for poor reproduction, the condition becomes worse because more females near *K* means fewer young (Fig. 37). If the population is harvested from *K* toward MSY on the *x* axis in Figure 37, however, recruitment of young is increased until MSY is reached—a counterintuitive outcome. If the objective is the maximum harvest the population can sustain, prudent management dictates that the population be backed away from a harvest at MSY to a fixed removal yield (FRY) to avoid an inadvertent overkill (Fig. 37). Unlike density-independent species, where the surplus determines the harvest (Leopold 1933), for ungulates, the harvest determines the surplus (McCullough 1979, Bowyer 2020b; Fig. 37). Harvest of most ungulate populations in North America is male-biased (Monteith et al. 2018, Bowyer et al. 2020b); consequently, a more substantial harvest of females would likely improve productivity of many populations. I suspect that most population ecologists understand the density-dependent relationships detailed herein. I wonder, however, how many know that these outcomes are a result of spatial segregation of the sexes.

Another characteristic of ungulate populations that can confuse interpretation of data is changing patterns of compensatory and additive morality (Errington 1946). Compensatory mortality is defined as one source of mortality compensating for another (e.g., animals killed during hunting season would have died anyway from harsh winter conditions or predation) (Bartmann et al. 1992, Boyce et al. 1999, Monteith et al. 2014b, Bowyer et al. 2020b). Addi-

tive mortality exists when sources of death are summed (e.g., animals killed by hunters would be added to those dying from other causes, which in the absence of harvest would not have died).

Amounts of additive and compensatory mortality vary with the dynamics of ungulate populations and the proximity of those populations to K (McCullough 1979, Bowyer et al. 2014, Monteith et al. 2014b, Bowyer 2020b). As an ungulate population increases from low density, where females are in excellent physical condition and are mostly successfully at recruiting young, to higher numbers near K, where females are in comparatively poor condition and less successful at provisioning young, a change occurs in the proportion of additive versus compensatory mortality. At those higher numbers, females attempt to rear more young than the environment can support. These young might perish from a variety of sources, but independent of the cause, they are destined to die—mortality is compensatory (Fig. 38). In some instances, compensatory mortality can result from seasonally determined processes of density dependence (Boyce et al. 1999). "Seasonality" is a consequence of the interaction between a reduction in population size and the density-dependent response, which results in more individuals birthed in the

Density-Dependent Recruitment

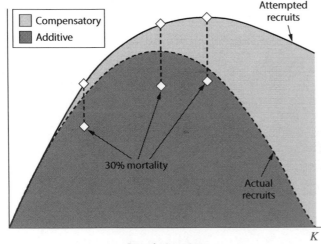

Fig. 38. Changes in number of successful recruits, as well as unsuccessful attempts to recruit, in relation to increasing size of an ungulate population. Females attempt to add more young to the population than can be sustained by the environment as a function of its carrying capacity (K). Mortality, which is largely additive at low density, becomes increasingly more compensatory as the population increases toward K. (Modified from Monteith et al. 2014b)

spring than can be successfully recruited through winter (Boyce et al. 1999).

Ungulate populations experiencing additive mortality tend to be at low density relative to K, with mortality becoming increasingly compensatory as population size and, thus, nutritional limitation increases (Monteith et al. 2014b, Bowyer et al. 2020b; Fig. 38). Conversely, a similar level of mortality becomes increasingly additive (sources of mortality are summed) as population size moves farther away from K, because the number of young that females attempt to recruit approaches the number of young they can recruit successfully given their nutritional status (i.e., most attempts at reproduction are successful) (Bowyer et al. 2020b; Fig. 38). Note that prescribing a harvest of 30% results in mortality that varies in its distribution between successful and attempted recruits as the population changes in size (Monteith et al. 2014b; Fig. 38).

These dynamics of ungulate populations are crucial for interpreting outcomes from predator-prey interactions. For example, if a deer population is near K, where principally compensatory mortality occurs, and suffers heavy mortality of young from coyote predation, there is little need for concern; predator control would not be warranted biologically and would do little to promote productivity of the deer population. Because mortality is compensatory, young deer would have perished from other causes had coyotes not killed them. In contrast, if the population is below MSY, where mortality is mostly additive, and coyotes kill a large number of young, concern over the productivity of the deer population is valid, and a biological justification for predator control would exist. Moreover, when ungulate populations are held at low density in predator pits (see Chapter 1 and Fig. 1), these same associations between additive and compensatory mortality pertain. I am not endorsing or opposing predator control—my interest is in providing an understanding of how predators and their management interact with the dynamics of ungulate populations, which are influenced by sexual segregation (see Fig. 4).

Circumstances exist where density dependence can be difficult to detect (Bowyer 2014, 2020b). In stochastic environments, forage availability may vary substantially among years (Mackie et al. 1990, Marshal et al. 2005, Heffelfinger et al. 2017). Consequently, the capacity of habitat to support ungulates can change annually, causing what could appear to be a lack of density dependence with density displaying no obvious relationship with nutritional or demographic variables (McCullough 1999). Density dependence, however, remains operative because resources available per capita are a result of both population abundance (i.e., density) and the availability of food within a particular year (McCullough 1990, Monteith et al. 2014b). Consequently, an apparent absence of a short-term relationship with population density does not necessarily imply an absence of density dependence over a greater time span (McCullough 1990, Kie et al. 2003).

Density-independent factors, at times, may seem to drive the dynamics of ungulate populations. One limitation to invoking weather (e.g., rain, snowfall, temperature) as a regulating mechanism for ungulates is that there can be an interplay of those environmental variables with population density of ungulates relative to K (Monteith et al. 2014b, Bowyer et al. 2020b). Ungulate populations near K will be in poor nutritional condition; weather is therefore likely to help or hinder those individuals disproportionately more than animals existing at lower numbers where their physical condition is excellent (Bowyer et al. 2014, 2020b). At high numbers relative to K, density dependence may outweigh even density-independent events that at lower densities might be helpful (Stewart et al. 2005). Moreover, effects of extreme drought may be overridden by increased nutrition related to reduced population numbers (McCullough 2001, Thalmann et al. 2015).

Spurious correlations between weather and vital rates, including pregnancy, young recruited, and survival, can exist for ungulates and may lead mistakenly to the conclusion that weather, rather than population size, is regulating a population, especially for populations near K (Bowyer et al. 2020b). Behavior and nutritional condition can help buffer ungulates against adverse effects of weather (Long et al. 2014). Moreover, the same weather conditions occurring at high density relative to K, which affect productivity of the population, likely would fail to do so at lower density when ungulates were in better physical condition. Compelling evidence exists of the effects of weather being more pronounced on moose populations experiencing resource limitations (Jesmer et al. 2021). Incorporating density-independent variables in models for ungulates based on such misconceptions can result in population models that cannot account for variation in population size relative to K (Bowyer et al. 2020b). Weather metrics, including winter severity, may be lurking variables (i.e., those that are correlated with the variable of interest, such as survival, but are not its primary cause), and their misinterpretation can lead to the mismanagement of populations. I acknowledge that on rare occasions weather can kill animals without regard to their physical condition (Bleich and Pierce 2001, O'Gara 2004a), but such events cannot be common or few animals would persist in such environments (Bowyer et al. 2020b). Nonetheless, assuming that density-independent variables are limiting an ungulate population without first examining the relationship of the population to K is not advisable.

I contend that the wise management and conservation of ungulate populations relies on one essential point—knowing where the female component of the population is relative to K. Estimating K, however, can be a difficult task. Regressing recruitment rate against population size can be used to estimate K, but that method often overestimates K, and it may require many years of data collection (McCullough 1979, Bowyer et al. 1999a). Forage-based methods (Hobbs and Swift 1985, Beck et al. 2006) for estimating K exist but are labor intensive; forage measurements also may lag declines of ungulate numbers. Other methods (Boyce 1989, Forsyth and Caley 2006) exist to

determine K but require large data sets and can be costly; years may be needed to obtain the information necessary to parameterize those models (Bowyer et al. 2005, 2013; Monteith et al. 2014a). Matters related to the conservation and management of ungulates often would be decided long before the aforementioned methods could be implemented (Bowyer et al. 2013). In addition, habitat or environmental changes might have occurred, potentially invalidating conclusions from the resulting models (Bowyer et al 2020b).

The harvesting of ungulate populations typically has a degree of uncertainty; biologists often must determine management objectives with limited information concerning the status of populations (Dinsmore and Johnson 2012). When more detailed and reliable data related to population dynamics are unavailable because of time or expense, varying the harvest of females and monitoring selected life-history characteristics and related metrics (Table 7) can resolve the status of a population and aid in determining whether the harvest is sustainable, leading to sound management decisions (Bowyer 2020b).

The life-history and population characteristics in Table 7 offer a conceptual framework with which to assess the nutritional status of ungulate populations and estimate their relationship to MSY and K. These characteristics change with the size of a population in relation to K, and they may help assess effects of harvest or predation. Moreover, ungulate populations exhibit a sequence of changes in life-history traits that tend to be altered as the population approaches K. The most sensitive of those characteristics is declining recruitment of young, followed by increasing age at first reproduction, declining litter size, lower rates of pregnancy, and finally, diminishing adult survival (Gaillard et al. 2000, Eberhardt 2002). I am not advocating where to manage an ungulate population relative to K. Such decisions are both socioeconomic and biological in nature, and they often vary with the objectives and responsibilities of the management agency. I do, however, provide the background to interpret the likely biological outcomes from such management decisions.

Table 7. Variation in life-history and population characteristics of ungulates in relation to the proximity of the population to MSY (maximum sustained yield) and *K* (ecological carrying capacity) (modified from Bowyer et al. 2014, 2020b)

Life-history and populations characteristics	≤MSY	Near *K*
Physical condition of adult females	Better	Poorer
Pregnancy rate of adult females	Higher	Lower
Pause in annual reproduction by adult females	Less likely	More likely
Yearlings pregnant[a]	Usually	Seldom
Corpora lutea counts of adult females[a]	Higher	Lower
Litter size[a]	Higher	Lower
Age at first reproduction for females	Younger	Older
Weight of neonates	Heavier	Lighter
Mortality of young	Additive	Compensatory
Diet quality	Higher	Lower
Population age structure	Younger	Older
Age at extensive tooth wear	Older	Younger

[a] Some species of ungulates may exhibit limited variability in particular characteristics.

There are several caveats to this approach of assessing where the population is relative to *K*. Maternal effects are widespread in cervids (Freeman et al. 2013) and might confuse interpretations using an approach based on some variables in Table 7, especially size of neonates. Nutrition of females is a primary determinant of male body and antler size; small males born to small females may never recover antler or body size, even on a high nutritional plane. Monteith et al. (2009) demonstrated that young males born to smaller females from the Black Hills of South Dakota remained smaller and had smaller antler sizes throughout their lives compared with young born to larger females from eastern South Dakota, despite all deer being fed the same highly nutritious diet (Fig. 39). Females required 2 generations to recover body size and give birth to males that ultimately had larger bodies (Fig. 39) and antlers (Monteith et al. 2009). These results indicate that habitat improvements or a reduction in females is critical for producing large males, but such management might not have an immediate effect on size of male offspring. In addition, harvesting small males to improve antler size will not have the desired effect in the absence of management that results in females being in good physical condition. Similarly, moose that were born small failed to compensate in size over the 10 months following birth, and size differences from birth were maintained over time (Keech et al. 1999). Multiple life-history and population characteristics in Table 7 should be used to evaluate where the population is with respect to *K*, but selecting which variables to measure may be unique to each study (Bowyer et al. 2020b).

Uncertainty in interpreting population dynamics of ungulates also may result from delayed density dependence, which occurs when recruitment is lagged further than expected following harvest or from other sources of mortality (Fryxell et al. 1991, Lande et al. 2006). Rangelands may take time to recover from overgrazing or other perturbations (Heady 1975); ungulate populations inhabiting those ranges may not respond quickly to harvest in the expected manner. Some diseases and parasites likewise can affect populations of ungulates (Cassirer and Sinclair 2007, Jones et al. 2017), although diseases may interact with reduced physical condition from high population density to amplify mortality (Sams et al. 1996). Providing that these caveats are kept in mind, the variables in Table 7 offer a useful method for judging where the population is relative to *K*, and thereby allowing assessments of harvest without necessitating estimates of absolute population size (Bowyer et al. 2020b).

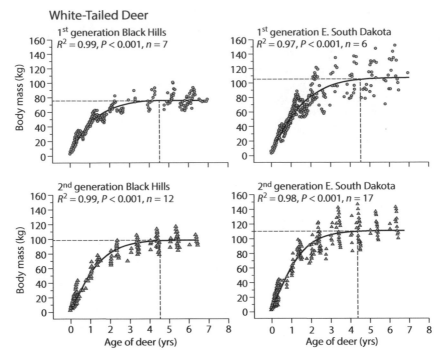

Fig. 39. Body mass (kg) relative to age (years) of male white-tailed deer raised in captivity in South Dakota; data were fitted with the von Bertalanffy growth curve. Deer were composed of original cohorts (first generation) acquired as neonates from eastern South Dakota and the Black Hills in southwestern South Dakota. Second-generation young were born in the research facility and were sired from first-generation adults. Sample size indicates number of individual deer and the dashed line represents time to asymptotic body mass. (From Monteith et al. 2009)

Adaptive management has been proposed as a method for coping with the presence of uncertainty in biological systems (Walters and Hilborn 1978, Westgate et al. 2013). Where more accurate and dependable data related to population dynamics are unavailable because of time or expense, varying the harvest of females and monitoring selected life-history characteristics and related metrics (Table 7) can help determine the status of a population and whether the harvest is sustainable. Also, the sequence in which those life-history traits respond can provide insights into where the population is relative to K (Bowyer et al. 2020b). Thus, in its simplest form, adaptive management involves a pattern wherein a harvest plan (or hypothesis) is developed and undertaken, and then results (from variables such as those in Table 7) are evaluated; learning occurs from interpreting those results, and the plan is adjusted taking those results into consideration. The process is then repeated as the harvest strategy is refined.

Some topics in this chapter may seem to have strayed somewhat from sexual segregation. My purpose, however, is to illustrate the pervasive nature of effects from sexual segregation on the ecology, behavior, and management of ungulates. I have suggested that the sexes of dimorphic ungulates be managed as if they were separate species. Harvest is a major component of most management plans for these large mammals, and understanding its consequences is not a simple process. Ecological linkages associated with sexual segregation are essential to appreciate their effect on ungulate population dynamics. Indeed, the downstream consequences from spatial segregation of the sexes permeate many life-history characteristics of ungulates, including the mating systems (Bowyer et al. 2020a) and population dynamics (Bowyer 2020b; see Fig. 4), which are linked inextricably to ungulate management and conservation.

SUMMARY

This chapter considers how sexual segregation in ungulates underpins important topics in evolutionary biology and ecology. Specifically, the role of sexual segregation in parental investment and population dynamics is elaborated. Ramifications from sexual segregation on those theories and their resulting consequences for management are discussed.

1. Sexual segregation in ungulates plays a major role in affecting patterns of parental care, competition among males for mates, and sexual selection. Spatial segregation of the sexes, especially around the time of parturition, limits paternal care; males have little opportunity to recognize their offspring, making male-male combat the primary mechanism for increasing reproductive success. Such male-male competition influences the evolution of sexual dimorphism and, consequently, the various mating systems exhibited by ungulates.

2. One outcome from spatial segregation of the sexes is that adult females compete more with one another and young than with adult males. Also, females occur at higher densities than males, which further intensifies competition among females. Hence, females are more involved in density-dependent processes and population regulation than males.

3. Ungulates exhibit strong density dependence with a decline in vital rates associated with increasing population size of females but not males. Density-dependence results in S-shaped growth curves that asymptote at K (environmental carrying capacity), which result from competition among females for food. Demographic characteristics that change with respect to the size of the population relative to K include survival of young, age at first reproduction, litter size, and weight of neonates.

4. Ungulate populations contrast with species exhibiting strong density independence. Those species tend to be small-bodied, have high reproductive rates, and have short lifespans; such species are limited by density-independent events, such as severe weather. Growth curves of density-independent species tend to be J shaped and typically do not reach K before being reduced by weather-related phenomena. The death of those animals constitutes the "doomed surplus," and those animals are available to harvest because they would die of other causes anyway (i.e., a harvestable surplus). Thus, the surplus determines the harvest.

5. Dynamics of ungulate populations are illustrated by a parabolic relationship between the number of animals recruited into the population and population size relative to K. The maximum harvest (or death from other causes) that the population can sustain occurs at the peak of the parabola, which is termed maximum sustained yield (MSY). The dynamics of ungulate populations are counterintuitive, and the harvest from such populations determines the surplus rather than vice versa. Thus, not harvesting females from populations at high density with respect to K may result in poor productivity, whereas harvesting them at densities near K increases productivity.

6. Additive and compensatory mortality also play important roles in the dynamics of ungulate population. Additive mortality (where sources of mortality are summed) tends to occur at low density relative to K, because females are in good physical conditions and most attempts at reproduction are successful. Compensatory mortality (where one source of mortality compensates for another) tends to occur when

populations are in poor condition near K, when females attempt to rear more young than the environment can support. This relationship is important for understanding the role of predation in regulating ungulate populations—predators will have a minimal effect on populations near K but can be more influential in populations at lower density where mortality is additive.

7. Changing patterns of compensatory and additive mortality can lead to misinterpreting effects of weather on mortality of ungulates. Populations near K are more likely to be helped or hindered by weather because they are in poor nutritional condition, whereas populations closer to MSY are less affected by the same weather conditions because they are in better physical condition. Correlations between weather and population characteristics of ungulates typically are not independent of populations size relative to K.

8. A suite of life-history and population characteristics can be used to determine the relationship between population size and K without having an estimate of population size.

9. Adaptive management wherein harvests of females is used to judge the sustainability of harvests in general can be an effective management strategy for ungulates

10. Sexual segregation has a profound effect on theories related to parental investment, sexual selection, mating systems, and population dynamics, with important consequences for ungulate management and conservation.

9 | Failing to Consider Sexual Segregation

Why is a knowledge of sexual segregation important? This chapter details management and conservation concerns for ungulates that lead to unanticipated outcomes because sexual segregation was not considered as a critical component in the ecology of these iconic mammals. Knowledge concerning sexual segregation in ungulates among individuals who manage natural resources should be expanded. Moreover, the failure to consider this critical component in the life history of ungulates can lead to inadvertent management errors, and can extend to serious problems related to understanding the ecology and behavior of the sexes of ungulates. This chapter contains abbreviated accounts of articles demonstrating such problems; more detailed descriptions and explanations are available in the original publications.

Moose

Habitat manipulations can inadvertently benefit one sex at the expense of the other. The first example is from a habitat manipulation performed to benefit moose in interior Alaska. In early to mid-March, a bulldozer was used to shear, crush, and break over old-growth feltleaf willow (*Salix alaxensis*) on about 200 ha of frozen ground adjacent to Goldstream Creek near Fairbanks, Alaska—the primary purpose

was to enhance forage abundance and quality for moose (Bowyer et al. 2001b); this manipulation also reduced concealment cover (Chapter 2). A near identical area immediately adjacent to the treated area was not crushed and served as a reference site. Tracks and feces indicated the study area was heavily used by moose during winter, although no estimates of population size were available. The moose population in the nearby Tanana Flats, however, had been increasing (Keech et al. 2000).

Responses of willows were quantified over 3 years (i.e., growing seasons) after the manipulation (Bowyer et al. 2001b). Willows were sampled on 10 quadrats (5 × 5 m) on both crushed and uncrushed areas. Number of leaders of current annual growth within reach of moose (≤3 m above the packed snow) and the number browsed by moose were tallied in each quadrat. The length of each leader of new growth was measured from the bud-scale scar to the terminal bud or to the point of browsing. Diameter of the leader at the bud-scale scar and the point of browsing also was measured. A sample of 15 leaders of current annual growth was clipped from each of the quadrats with willows.

In the laboratory, leaders clipped in the field were measured and then dried to a constant mass, dry weight of leaders was calculated, and leaders from each quadrat were pooled into a composite sample for analysis. Measures of forage quality, including acid-detergent fiber, neutral-detergent fiber, acid-insoluble ash, lignin, hemicellulose, cellulose, tannins, nitrogen, and in vitro dry-matter digestibility, were obtained with standard procedures.

Moose tracks were sampled along 10 randomly located transects, each 25 m in length in both crushed and uncrushed areas. Each transect was oriented either north-south or east-west based on a coin flip. Fresh tracks of moose crossing transects in either direction were recorded. Snow depth along transects was also recorded (Bowyer et al. 2001b). Fecal pellets of moose were collected while sampling vegetation and during track counts. In addition, a systematic search was conducted on both crushed and uncrushed areas for fecal groups; this was repeated 1 week later. The mean volume (mm³; i.e., length × width × depth) of 5 pellets from each fecal group was determined. Computations from Mac-Cracken and Van Ballenberghe (1987) were then used to estimate the sex and age class of moose that deposited a particular pellet group.

Mean depth of snow (~43 cm) was similar on crushed and uncrushed sites. Mean number of moose tracks per transect was also similar between areas (uncrushed = 1.9 tracks; crushed = 1.6 tracks). In addition, size of leaders of current annual growth within the reach of moose did not differ between crushed and uncrushed areas. Most leaders of new growth on the uncrushed areas were stump sprouts growing from the trunks of old willows and were similar in architecture to leaders regrowing from crushed plants (Bowyer et al. 2001b). Nonetheless, crushing of willows resulted in nearly 5 times more leaders of current annual growth available to moose in the crushed area compared with the uncrushed area. Similarly, dry mass of total leaders was >3-fold higher on the crushed than the uncrushed area. Moose browsed more leaders on the crushed (62%) compared with the uncrushed (28%) area. Clearly, crushing provided substantially more forage for moose.

Surprisingly, few differences existed among a suite of variables related to the nutritional quality of willows between crushed and uncrushed areas (Bowyer et al. 2001a). Overall low values for nitrogen (~0.8%) and in vitro dry-matter digestibility (~30%) indicated that current annual growth of willows provided relatively low-quality forage; nonetheless, willows are preferred forage for moose during winter (Weixelman et al. 1998).

Unexpectedly, volume of fecal pellets of moose collected on crushed and uncrushed areas indicated that male moose used the crushed area proportionally more often than females and young, and the uncrushed areas were used less often (Fig. 40). The stand of ~60-year-old willows offered substantial concealment cover compared with the open features of the crushed area. A plausible explanation for females and their young seeking such areas was to lower predation risk, whereas males took advantage of the more abundant forage resulting from

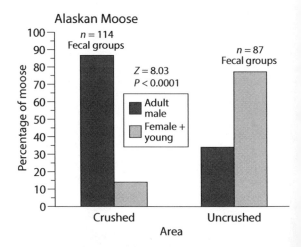

Fig. 40.   Percentage of adult male and adult female plus young Alaskan moose occurring on areas where willows were mechanically crushed 3 years earlier, and on an adjacent area with an old-growth stand of willows (uncrushed) in interior Alaska during winter. Sex and age classes of moose were determined from volume of fecal pellets (MacCracken and Van Ballenberghe 1987). The *Z* test and associated *P* values are from the comparison of the proportion of adult males occurring on crushed and uncrushed areas. (From Bowyer et al. 2001b; originally published in *Alces*)

the crushing of willows. Consequently, management designed to improve winter habitat for moose inadvertently lowered the suitability of the habitat for females and young while improving habitat for males.

North American Elk

The second case study illustrates how habitat manipulations for other purposes can affect habitat selection by the sexes differentially. Responses of male and female North American elk to a fuels-reduction program at the Starkey Experimental Forest and Range (hereafter Starkey) in the Blue Mountains of Oregon were studied by Long et al. (2009b). A spruce budworm (*Choristoneura occidentalis*) outbreak in the late 1980s had led to extensive mortality of true fir (*Abies* spp.) and Douglas fir (*Pseudotsuga menziesii*) throughout Starkey, which resulted in high loads of dead and standing fuels and an increased likelihood of wildfires. Twenty-six stands, encompassing about 9% of the study area, were randomly chosen for treatment with mechanical thinning followed by prescribed burning across 3 years; 27 comparable stands were left untreated to serve as reference areas. Treatment stands were thinned mechanically between May and October and then treated with prescribed fire during September or October of either the same year or the following year. From 90% to 100% of the area of each treatment stand was broadcast burned with low- to moderate-intensity fires. Stands were assigned to 1 of 5 categories: control, 2-year-old burn, 3-year-old burn, 4-year-old burn, and 5-year-old burn (Long et al. 2009b).

Locations of radio-collared adult male and adult female elk were studied in spring and summer during periods of activity (Long et al. 2009b). Use of treatment stands by male elk during active periods in spring was lower than use of those stands by female elk relative to availability (i.e., male elk avoided [selected against] treatment stands). Patterns of stand use by the sexes typically were more similar during summer than spring, but the direction of differences

in selection ratios (i.e., selection or avoidance) between females and males was less consistent.

Female elk either avoided treatment stands (3- and 4-year-old burns) or used them in proportion to their availability (2- and 5-year-old burns) in summer. A similar pattern was evident for males, but males avoided 2- and 5-year-old burns and used 3- and 4-year-old burns in proportion to their availability. Mean selection ratios for control stands were near equal between sexes and indicative of mild selection of those stands by both sexes during summer. Females used control stands significantly more often than 3- and 4-year-old burns relative to their availabilities in summer, an opposite pattern to selection by females in spring. Nonetheless, males used control stands significantly more than 2- and 5-year-old burns in summer relative to their availabilities, a pattern similar to stand use by males during spring (Long et al. 2009b).

Overlap in space between sexes of elk during times of peak activity was higher in summer than spring (Long et al. 2009b; Fig. 41). Nevertheless, spatial overlap of same-sex groups (female-female and male-male) also increased during summer (Fig. 41). In both seasons, spatial overlap was significantly higher among females than between females and males (Fig. 41). Mean overlap among males was intermediate between means for the other 2 groups during both seasons, but it did not differ significantly from means for those groups in either season (Fig. 41).

Long et al. (2009b) concluded that habitat manipulation at Starkey influenced patterns of space use by female and male elk, and that fuels reduction likely affected the degree of sexual segregation. These observations add support to the hypothesis that differential selection of foraging habitat by the sexes plays an important role in influencing sexual segregation. These results also hold important implications for forest and wildlife managers. The interaction between fuels reduction and seasonal changes in plant phenology at Starkey resulted in a mixture of burned and unburned forest habitats

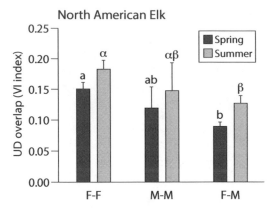

Fig. 41. Mean volume of intersection (VI) index values and 90% confidence intervals within and between sexes of North American elk in spring (1 April–14 June) and summer (15 June–31 August) at the Starkey Experimental Forest and Range, Oregon. VI index values indicate the degree of overlap in the volumes of two 99% fixed-kernel utilization distributions (UDs) and ranges from 0 to 1, with 0 indicating no overlap and 1 indicating complete overlap. Means were calculated from all possible pairwise comparisons of UDs within groups and seasons. Unshared letters among groups within each season (Latin for spring, Greek for summer) indicate a significant difference between means. Group abbreviations are: F-F = female-female comparisons, M-M = male-male comparisons, and F-M = female-male comparisons. (From Long et al. 2009b)

that likely provide better long-term foraging opportunities for female elk than what burning a large proportion of a landscape would provide. Such a strategy also may minimize negative effects of fuels reduction on male elk, especially for burns of a comparatively young age (≤5 years old). Nonetheless, these results provide additional evidence that manipulation of some habitats benefited females, albeit to the potential detriment of males.

White-Tailed Deer

The previous studies are among the few that explicitly consider whether manipulation of habitat to benefit ungulates might result in males and females responding differently to alterations of their environment. Stewart et al. (2003b) examined effects of habitat manipulation on sexual segregation in white-tailed

deer in South Texas by creating clearings within a shrubland matrix that altered habitat and forage.

Stewart et al. (2003b) conducted research on the Rob and Bessie Welder Wildlife Foundation Refuge located about 11 km north of Sinton, Texas. Vegetation in treatment areas included 2 shrubland plant communities, which contained a mixed-grass chaparral and live-oak (*Quercus virginiana*) chaparral. Stewart et al. (2003b) created clearings in continuous shrubland by mechanically removing woody plants to ground level using roller choppers pulled by a crawler tractor during mid-June. They then used chemical herbicides to alter availability of forbs and shrubs within those open patches. Treatments were applied at a scale of 4 ha. Treatment applications created a mosaic of openings with varying availabilities of forages surrounded by mature shrubland. The vegetative mosaic surrounding each block was composed of about 150 ha in mixed-grass chaparral (2 blocks) and about 100 ha in the live-oak chaparral (1 block). Treatment applications of herbicides and mechanical reduction of shrubs resulted in 5 categories related to the availability of forbs, grasses, and shrubs.

Stewart et al. (2003b) observed white-tailed deer daily from August to December, when deer were aggregated in mixed-sex groups, and from March to July, when sexual segregation occurred. Aggregation included rut, and segregation encompassed parturition. Observations were made during crepuscular periods to encompass peaks of deer activity. Deer were observed in each treatment with binoculars from a 4-m-tall stand and counted by scan sampling at 15-min intervals (Altmann 1974). Treatments were selected randomly within a block (without replacement) for observations of deer activity. Each treatment was observed for an evening and the following morning. Stewart et al. (2003b) then shifted to the next randomly selected treatment until all treatments in a block were observed prior to moving to the next block. Deer were observed for 4 rotations of each treatment-block combination for each season, including 120 observation periods (240 hr) each during sexual segregation and sexual aggregation.

Adult female white-tailed deer did not differ in their use of treatments among seasons, and they did not exhibit a season-by-treatment interaction. Thus, seasons were combined for that analysis. Females differed significantly in use of treatments—chop only was used more by adult females than the reference area (control) or the treatment that reduced forbs and shrubs. Adult males differed significantly in use of treatments among seasons, with a significant season-by-treatment interaction. During autumn, males differed in use of cleared treatments and used chop only to a greater extent than other treatments. During spring, males did not differ in their use of control or other treatments (Fig. 42). Clear differences existed in the manner in which the sexes of white tailed deer responded to various manipulations of habitat.

Stewart et al. (2003b) documented sexual segregation in white-tailed deer at a scale of 4 ha. The researchers successfully manipulated forage availability in cleared treatments, particularly in the treatment that reduced forbs and shrubs, which also resulted in an increase in biomass of grasses. Reduction of shrub diversity in areas where forbs are not available or during seasons when forbs are reduced likely had a disproportionate effect on female versus male white-tailed deer. Indeed, adult females consistently used the treatment that reduced both forbs and shrubs to a lesser extent than the chop-only treatment during both seasons, illustrating how habitat manipulations and resulting forage differences affected the distribution of male and female white-tailed deer.

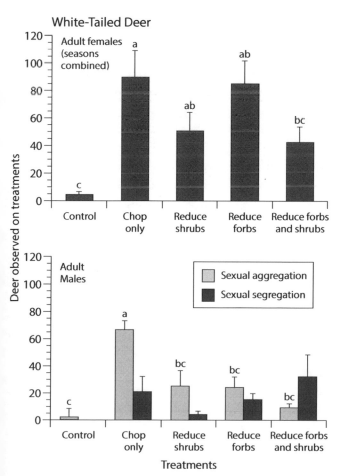

Fig. 42. Mean number of white-tailed deer (1 SE) using treatments, by gender, on the Rob and Bessie Welder Wildlife Refuge in South Texas. Letters above bars indicate results of contrasts from repeated measures analysis of variance comparing treatment differences, in which different letters indicate differences ($P<0.03$). (From Stewart et al. 2003b)

Bighorn Sheep

Implementation of habitat modifications to benefit a species without considering patterns of sexual segregation can lead to less than optimum results. Whiting et al. (2010) investigated use of water sources by the sexes of bighorn sheep on Antelope Island State Park in Utah. Water sources are important for the persistence of populations of bighorn sheep, especially in arid environments. Moreover, understanding whether the sexes use different water sources holds importance for the conservation of this species and the surrounding habitats they occupy.

Whiting et al. (2010) defined seasons according to birthing and mating behaviors of bighorns on the island: spring was 1 April to 30 June, summer consisted of 1 July to 30 September, and rut occurred from 1 October to 30 November. Winter (1 December to 31 March) was not included because bighorns rarely visited water sources during that time. Sexual segregation was defined as 1 April to 30 September, which encompassed parturition, and sexual aggregation as 1 October to 30 November based on mating behaviors of these bighorn sheep.

Whiting et al. (2010) deployed motion-sensor cameras at 7 natural sources of water, which were the only known perennial sources of water in proximity to bighorn sheep habitat and ranged in elevation from 1,290 to 1,680 m. Pictures were taken every 20 sec while animals were in the field of view of the sensor of the camera. Following a lapse of 25 min after a bighorn sheep activated a camera and was recorded, a camera activated by a bighorn was scored as a new visit. With this criterion, the median number of hours between successive visits to a water source was 13 (upper and lower quartile distances = 2–25 hr).

Those authors observed bighorn sheep an average of 3 times each month from April to November 2005 and 2006. They conducted ground observations during daylight in areas used by these ungulates. Bighorn sheep were observed from designated trails and game trails leading to sources of water in bighorn sheep habitat. When groups of bighorn sheep were observed, sex of individuals and group size and composition were noted. Undisturbed animals were considered a part of the same group if they were ≤50 m from one another, or if they appeared to be aware of other bighorn sheep and moved as a cohesive group. Each location was mapped, and 50% core areas were calculated for males and females during segregation and aggregation with adaptive-kernel home ranges using home-range tools (Rodgers et al. 2007). Overlapping use of space between sexes was quantified by calculating a utilization distribution overlap index (UDOI) (Fieberg and Kochanny 2005) between 50% core areas of males and females (Whiting et al. 2010).

Cameras were set for 258 days when sexes of bighorn sheep sexually segregated and 122 days when they sexually aggregated. During segregation, males visited water sources A through D on 39 occasions and E through G on 205 occasions, whereas females visited springs A through D on 329 occasions and E through G on 46 occasions (Fig. 43). Thus, water sources E through G were used mostly by males, and springs A through D were used predominantly by females during sexual segregation (Fig. 43). During aggregation, males visited water sources A through D on 77 occasions and E through G on 17 instances, whereas females visited springs A through D on 30 occasions and E through G on only 7 instances (Fig. 43).

Whiting et al. (2010) observed groups of males on 82 occasions and groups of females on 140 occasions during segregation, and groups of males and females on 38 occasions during aggregation; those data were used to calculate 50% core areas. During segregation, the 50% core area used by males (3.9 km²) was 3 times as large as the core area used by females (1.2 km²), and the size of the 50% core area used by sexes during aggregation was 8.1 km² (Fig. 43). During segregation, 3 water sources were outside of the core area used by males, and all sources of water were outside the core area used by females (Fig. 43). During aggregation, 2 sources of water were outside the

Bighorn Sheep

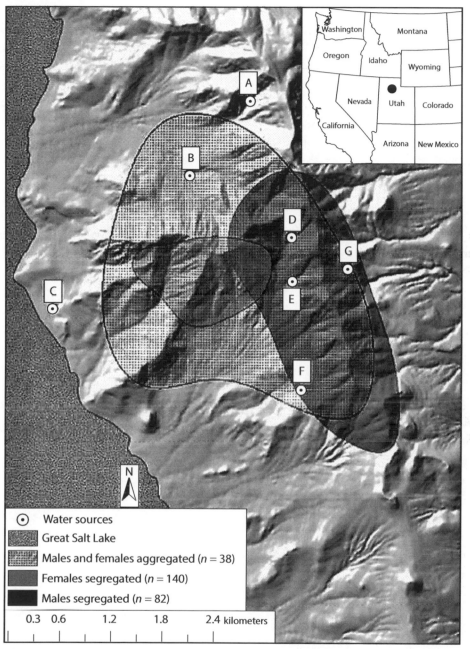

Fig. 43. Utilization distributions (50% core areas), and number of sightings used to produce polygons, in relation to sources of water (designated A–G) for male and female bighorn sheep during segregation and aggregation on Antelope Island State Park, Utah, over 1.5 years. (From Whiting et al. 2010)

core area used by both sexes (Fig. 43). The UDOI was 0.023, reflecting a small degree of overlap in volume of 50% core areas of males and females.

When sexually segregated, male and female bighorn sheep used different sources of water on Antelope Island (Whiting et al. 2010). During aggregation, males visited water sources used by females almost 3 times more often than when sexes segregated, and both sexes of bighorns used water sources more during summer compared with other seasons. The authors concluded that a need exists to assess the availability of water sources near habitat used by males and females before conserving and manipulating habitat, siting artificial sources of water, or reintroducing bighorn sheep. Similar results regarding differential use of artificial water sources by the sexes during sexual segregation have been documented for mule deer inhabiting an arid environment (Shields et al. 2012).

Human disturbance and modifications of habitat also hold potential to affect sexual segregation in ungulates (Rubin and Bleich 2005). For instance, anthropogenic developments can affect migration in mule deer (Lendrum et al. 2012, 2013). Further, development of a recreational hiking trail in North Dakota resulted in bighorn sheep being displaced from, and ultimately abandoning, a critical birthing area (Wiedmann and Bleich 2014). Clearly, such developments and activities hold potential to affect sexual segregation, but seldom have they been adequately addressed.

Habitat fragmentation and barriers to movement are a detriment to the long-term conservation of ungulates and other large mammals (Berger 2004). Roadways are a known cause of loss, fragmentation, and degradation of habitat (Coffin 2007). Bighorn sheep are notably affected by roadways and may experience increased mortality from collisions with automobiles (Rubin et al. 2002) or a reduction in gene flow and a decline in genetic diversity resulting from restricted movements (Epps et al. 2005).

In this second account for bighorn sheep, Bleich et al. (2016) investigated the propensity of the sexes to cross a 2-lane paved road and a single-lane main-

tained dirt road during periods of sexual segregation and sexual aggregation to assess whether those roadways affected movements of those ungulates. Research was conducted in eastern San Bernardino County, California, in the vicinity of Old Dad Peak in the Mojave Desert.

The study site was bisected on an east-west basis by Kelbaker Road, a 2-lane paved road that was lightly traveled (~100 vehicles per day). The study area also was divided on a north-south basis by a single-lane graded dirt road. Vehicle use on the dirt road was ≤1 vehicle per day. Bighorn sheep exhibited sexual segregation from December to July, and sexual aggregation from August to November (Bleich et al. 1997). Adult males moved from mountain ranges within the study area to join females for mating (sexual aggregation), and then separated from females during the birthing season (sexual segregation). Those movements to and from female ranges required that males cross the paved road at least twice annually. Female ranges were bisected by the single-lane dirt route. Neither female nor male bighorn could avoid effects of Kelbaker Road or the dirt road completely because those routes bisected the entire study site.

Bighorn sheep were captured with a handheld net gun fired from a helicopter. Sheep were fitted with very high frequency (VHF) radio collars with a 6-hr delay mortality sensor. Females ≥2 years old were considered adults; age categories of males were combined to obtain an adequate sample size (Bleich et al. 2016). A fixed-wing aircraft with an H antenna on each wing strut was used to locate individuals on a weekly basis from October to December for 5 years.

Bleich et al. (2016) used locations of individual bighorn sheep to determine the minimum number of times that individuals crossed Kelbaker Road or the dirt road during periods of sexual segregation and sexual aggregation. The size of core areas used by bighorn sheep was determined with the home-range analysis and estimation software HoRAE, and a 50% volume kernel-density estimate was calculated using the standard bandwidth, standard sextant biweight kernel, and a 25-m grid-cell size (Steiniger and Hunter 2012). The degree of overlap among 50% core areas was estimated with the UDOI (Fieberg and Kochanny 2005). UDOI values were calculated using the kernel-overlap module in the package "ADEHABITAT" for the R software (Calenge 2006).

On an annual basis, Bleich et al. (2016) reported that a significantly greater proportion of adult male bighorn sheep crossed Kelbaker Road than adult females (Table 8). Further, a significant difference occurred between proportion of males and females crossing Kelbaker Road during sexual segregation (December–July). No difference between the sexes occurred, however, in the proportion of animals crossing Kelbaker Road during sexual aggregation (August–November; Table 8). Similarly, there were no differences in the number of times males and females crossed Kelbaker Road during either sexual segregation or sexual aggregation (Table 8).

On an annual basis, a significantly greater proportion of male bighorn sheep crossed the dirt road that females (Bleich et al. 2016; Table 8). During sexual segregation and aggregation, significant differences existed in the proportion of males and females crossing the dirt route (Table 8). No differences occurred in the number of times males or females crossed the same route during either segregation or aggregation (Table 8). The sexes exhibited few differences in crossing the paved Kelbaker Road or the graded dirt road (Bleich et al. 2016).

The lowest overlap in 50% core areas was between male and female bighorn sheep during sexual segregation. Not unexpectedly, the greatest overlap in the 50% core of the home ranges for the sexes was during sexual aggregation (Bleich et al. 2016). Males tended to move to and from female areas during sexual aggregation and segregation, respectively, sometimes over distances of $\geq$15 km (Bleich et al. 1997).

Roadways may affect ungulates in at least 2 important ways. First, ungulates make decisions concerning the locations of home ranges by considering landscape features that lie outside of their home range (Kie et al. 2002). Thus, ungulates may select against habitat adjacent to roadways, even if the roadway is

Table 8. Predictions and outcomes for the potential effects of roadways on sexual segregation of 25 male and 17 female bighorn sheep annually as well as during sexual segregation (December to July) and aggregation (August to November). Significant differences are in bold (from Bleich et al. 2016).

Feature	Response by sex		
	Male	Female	*P* value
Kelbaker Road			
Proportion of crossings (annual)	56%	29%	**0.037**[a]
Proportion of crossings (segregation)	48%	24%	**0.044**[a]
Proportion of crossings (aggregation)	44%	29%	0.174
$\bar{x}$ (and *CV*) number of crossings (annual)	3.2 (118)	2.1 (208)	0.170
$\bar{x}$ number of crossings (segregation)[b]	0.21	0.16	0.116
$\bar{x}$ number of crossings (aggregation)[b]	0.39	0.18	0.131
Unnamed dirt route			
Proportion of crossings (annual)	85%	41%	**<0.001**[a]
Proportion of crossings (segregation)	72%	35%	**0.007**[a]
Proportion of crossings (aggregation)	68%	41%	**0.044**[a]
$\bar{x}$ (and *CV*) number of crossings (annual)	6.2 (101)	4.9 (142)	0.080
$\bar{x}$ number of crossings (segregation)[b]	0.50	0.43	0.116
$\bar{x}$ number of crossings (aggregation)[b]	0.54	0.44	0.171

[a] Values were statistically significant at *P* < 0.05.

[b] Values were corrected for the number of months during periods of segregation and aggregation.

not included in the home range, thereby influencing patterns of resource selection (Bleich et al. 2016). Second, ungulates may be forced to cross roadways to engage in necessary behaviors, such as foraging and reproduction, and that consequently increases the probability of them being struck by vehicles. Roads pose threats to ungulates via direct mortality, especially when speed limits are high, and also by affecting habitat selection, and potentially gene flow (Bleich et al. 2016). Bleich et al. (2016) strongly recommended that the potential effects of anthropogenetic developments on ungulates relative to their patterns of sexual segregation and aggregation be considered when implementing management plans, especially for development or modification of roadways.

Examples in the published literature of how agencies and others are managing the sexes of ungulates differently are nil, although I expect some unpublished management plans exist. Nonetheless, the potential for mismanagement is enormous when the importance of sexual segregation has been underappreciated or unrecognized. A pressing need exists to manage the species of dimorphic ungulates as if they were separate species; this includes the development of separate management plans and harvest strategies for the sexes.

SUMMARY

This chapter provides examples where management activities or anthropogenetic developments have interacted with patterns of sexual segregation to adversely affect the management or conservation of ungulates. Some actions inadvertently benefited one sex at the expense of the other. Moreover, management of some essential resources has failed to consider the distribution and differential needs of males and females while sexually segregated. Potential effects of developments, such as roadways, on the movements of ungulates moving to and from areas of segregation and aggregation have rarely been considered. Sexual segregation should not be ignored in the management and conservation of dimorphic ruminants.

1. Crushing old stands of willow to improve habitat for Alaskan moose during winter resulted in increased forage on the crushed area compared with the uncrushed area. Males benefited from the more open crushed areas, while females and young remained in the areas with better concealment cover.

2. Responses of male and female North American elk to a fuels-reduction program, which involved the crushing and burning of dead fir trees, were studied in the Blue Mountains of Oregon. A comparison of treated and control stands during sexual segregation indicated that females likely benefited from the fuels-reduction program but males did not.

3. White-tailed deer responses to clearings in shrublands were studied in South Texas during periods of sexual segregation and sexual aggregation. Various treatments involving mechanical brush removal and herbicides were applied to create a matrix of patches in the shrublands. Clear differences existed in the manner in which the sexes of white-tailed deer used various manipulations of habitat during periods of segregation and aggregation. This experiment indicated that reduction of shrub diversity in areas where forbs are not available, or during seasons when forbs are reduced, may have a disproportionate effect on female versus male white-tailed deer. Adult females consistently used the treatment that reduced both forbs and shrubs to a lesser extent than the chop-only treatment during sexual segregation and aggregation.

4. Bighorn sheep were studied on Antelope Island, Utah. When sexually segregated, males and females used different sources of water. During sexual aggregation, males visited water sources used by females nearly 3 times more often than when sexes were segregated; the sexes of bighorn sheep used water sources more during summer compared with other seasons. During segregation, 3 water sources were outside of the core

area used by males, and all sources of water were outside the core area used by females. During aggregation, 2 sources of water were outside the core area used by both sexes. A clear need exists to assess the availability of water sources near habitat used by males and females before conserving and manipulating habitat, siting artificial sources of water, or reintroducing bighorn sheep.

5. The propensity of the sexes of bighorn sheep to cross a 2-lane paved road and a single-lane maintained dirt road during segregation and aggregation was studied in the vicinity of Old Dad Peak in the Mojave Desert of California.

Ungulates may select against habitat adjacent to roadways, even if the road is not included in the home range, thereby influencing patterns of resource selection. Ungulates may be forced to cross roadways as they move to and from areas used during sexual aggregation and segregation, consequently exposing them to risk of collisions with vehicles. Potential effects of anthropogenetic developments on ungulates relative to patterns of sexual segregation and aggregation should be considered when implementing management plans, especially for development or modification of roadways.

10 | The Future

Defining Sexual Segregation

The most fundamental problem affecting our understanding of sexual segregation and its consequences for management and conservation is the lack of an operational definition of sexual segregation (Barboza and Bowyer 2000). This oversight has led to a hodgepodge of variables that are potentially necessary to measure to detect sexual segregation (see Table 3). The absence of a generally accepted definition for sexual segregation has resulted in confusion, and knowledge has been pursued down a number of unproductive pathways. Rectifying this issue is the first step in moving forward scientifically.

When it was thought that social segregation had the ability to explain differences in the use of space by the sexes (Conradt 1998a, 1998b; Ruckstuhl 1998, 1999, 2007), dividing sexual segregation into social and ecological components seemed reasonable (Mysterud 2000)—that conclusion is no longer valid. None of the hypotheses for social segregation can explain why the sexes of dimorphic ungulates spatially segregate into widely spaced areas (Bleich et al. 1997, Schaller 1998; Chapters 5 and 6). Moreover, I maintain that social hypotheses concerning sexual segregation relate primarily to the evolution of group living, including hypotheses involving social preferences, aggression between the sexes, and activity patterns (Chapters 5 and 6). Those hypotheses cannot explain independently why the sexes segregate and aggregate on a seasonal basis. I argue that some of those social hypotheses, especially those relying on differences in activity patterns, are better considered under a social-constraints hypothesis (Chapter 6). Those social hypotheses offer new insights into factors constraining the evolution of social groups but fail to explain spatial segregation of the sexes. Moreover, social hypotheses have limited value for dealing with the management and conservation of ungulates (Stewart et al. 2011).

Ecological and social segregation are completely different evolutionary processes and cannot be compared or assessed together to understand spatial segregation of the sexes. The methods of Conradt (1998a, 1998b) for differentiating social and habitat segregation are mathematically sophisticated but biologically meaningless. The idea of comparing apples to oranges applies in this relationship. Moreover, too narrow of a definition for habitat often is used in such comparisons, usually just referring to a vegetation type; the definition of habitat as it relates to an animal's niche is far broader (Hall et al. 1997). Nonetheless, the continuing use of such comparisons is evident (Crampe et al. 2021), despite those and other shortcomings with this method. The experimental research of Kie and Bowyer (1999) clearly demonstrated that many aspects of social behavior are uncoupled from the degree of spatial segregation, a point that has not be addressed by those supporting social underpinnings for sexual segregation. I contend that the traditional definition is still the most germane for understanding sexual segregation—the differential use of space or other resources by the sexes outside the mating system. Even with acceptance of that definition, a number of factors still need to be considered, including a niche-based approach that incorporates space, habitat, and food to understanding sexual segregation (see Table 5), definitions for groups (see Fig. 16), effects of scale (see Fig. 26), and potential tradeoffs between resources and predation (see Fig. 35). I further maintain that there may be no more than 2 general explanations for sexual segregation: resources (e.g., the gastrocentric hypothesis) and risk of predation (Chapters 2 and 7). Refining and refocusing our approach to understanding the causes of sexual segregation in ungulates is essential for the conservation and management of these iconic mammals. I, along with my students and colleagues, have long-maintained that the sexes of ungulates should be managed as if they were separate species (Kie and Bowyer 1999; Stewart et al. 2003b; Bowyer 2004; Bowyer and Kie 2004; Schroeder et al. 2010; Whiting et al. 2010; Oehlers et al. 2011; Bowyer et al. 2014, 2020b; Stewart et al. 2015). Unfortunately, I fear this is not a widely held view among resource professionals.

Why Is Sexual Segregation Overlooked?

A knowledge that sexual segregation occurs in ungulates is not new. Over a century ago Boner (1861) described this behavior for red deer. Since then, there have been >150 publications on this topic for ungulates alone (see Table 1), and those publications have increased in numbers over time (Bowyer 2004). Perhaps one underlying reason why sexual segregation is poorly understood is that it does not fit neatly into a single scientific discipline but straddles ecology and behavior. Moreover, information on sexual segregation seldom finds its way in to basic texts on either ecology or behavior (I failed to find any). Likewise, this topic is lacking in modern and otherwise excellent texts on mammalogy. I am less certain about textbooks on ornithology and ichthyology, but sexual segregation was not a topic of discussion in any of the classes I took on those subjects as an undergraduate.

In addition to being a mammalogist, I am also a wildlife biologist. Most texts in wildlife ecology and management, or conservation biology, likewise, do not list sexual segregation in their indices, including the recent edition of the *Wildlife Techniques Manual* (Silvy 2020). Notably, 2 textbooks on wildlife management and conservation do deal with sexual segregation (Krausman 2002, Krausman and Cain 2013), which I hope will set the standard for the future. How can biologists and resource managers deal with the consequences of sexual segregation if they do not know it exists? Sexual segregation and its role in conservation and management would become better appreciated with a more well-rounded approach to this topic. Textbooks used to educate students in the life sciences should include information on sexual segregation. Of course, there must be an interest in and knowledge of sexual segregation for those teaching such courses. Sexual segregation has wide-ranging effects on the ecology and evolution of ungulates (Chapters 8 and 9) and other species (see Table 1), yet this topic has had difficulty in finding its way into mainstream biology.

Undergraduate curricula undertaken by many students limits their exposure to the concept of sexual segregation. Many of the building blocks of an undergraduate education in biology involve STEM (science, technology, engineering, and mathematics) courses, including basic biology, chemistry, physics, and mathematics. Recently, E. O. Wilson discussed some of the problems he believed were inherent in a STEM education in *The Chronicle of Higher Education* (interview with C. Tyson, 7 May 2019). I concur with many of his criticisms.

Some students perish under a heavy dose of science before reaching the upper-division courses with the content that initially prompted their interest in obtaining a university education. For instance, organisms discussed in mammalogy, ornithology, herpetology, ichthyology, entomology, and other "-ology" classes are not reached until the later portion of bachelor (BA or BS) degrees. This pattern holds true for ecology, conservation, and wildlife management. These courses are typically upper division and require a strong background in science to cope with the material they contain. But, in this process, students may lose an appreciation for natural history, which often was the reason for selecting various majors in the biological sciences. Moreover, the opportunity to apply knowledge concerning the natural history of organisms to upper-division offerings is lost without at least some basic background on this topic. Indeed, Bleich and Oehler (2007) highlighted the need for training in natural history for those pursuing degrees in wildlife conservation. I have long believed that an introductory course in natural history, scattered with a background in Darwinian evolution, should complement STEM coursework. Such a course could offer a more exciting pathway to the upper division for many students, hold their interests, and better prepare them (including those with molecular or physiological bents) for a changing world. Some of these undergraduates will become the ecologists, conservation biologists, and wildlife managers of the future, and they will be responsible for the wellbeing of our environment. I hope that an understanding of why the sexes of animals use space differently will be part of their education.

I anticipate that the importance of sexual segregation in the conservation and management of animals will become more fully appreciated—I fervently hope this book aids in that process. Bennett et al. (2019) recently provided a persuasive article on the relevance of sexual segregation to the conservation of migratory land birds. Moreover, Santora (2020) wrote an excellent popular article in *Scientific American* discussing the degree to which sexual segregation has been overlook in the management of animals. Nonetheless, agencies charged with fish, wildlife, and natural-resources management and conservation have yet to fully incorporate the spatial separation of the sexes into their management plans. I argue that males and females of dimorphic ungulates should have separate management plans for both harvest and habitat modifications—this likely holds for other taxa as well. Indeed, sexual segregation is widespread among organisms (see Table 1). Where such information is lacking, those gaps in knowledge should be high priorities for additional research.

A number of new and exciting techniques exist that can be applied to questions concerning sexual segregation (Chapter 4), and there have been recent advances in our understanding of ungulate ecology (Bowyer et al. 2021). Observations of animals can be fascinating—often yielding results that are difficult to obtain with other methods. Thus, future research should not completely tradeoff new technologies for observational studies that help build an appreciation for and understanding of nature. Field approaches can complement modern techniques and broaden our understanding of ungulates. No technical innovation or metadata set can completely compensate for careful observations of animals under natural conditions. I believe that coupling fieldwork with modern analytical techniques offers the most promising route for resolving questions concerning sexual segregation.

Sexual segregation lies at the interface of theoretical and applied topics in ecology, behavior, and conservation. When funding is limited and the need for information urgent, a tendency exists to prioritize applied over theoretical investigations or action plans. Nevertheless, overlooking sexual segregation may result in harvest regimes that will not provide expected outcomes (Chapter 8), or result in unintended mismanagement of resources (Chapter 9). I contend there is a need to maintain a balance between theoretical and applied research. Most advances in applied biology have their origins in theoretical endeavors. Sexual segregation is a classic example of what was thought to be a purely a theoretical undertaking slowly becoming an essential element in bolstering applied efforts. Depending on the situation and the answers needed, the allocation of resources to theory and application need not be equal, but at least some of each should be maintained—outcomes from theoretical research have often provided unexpected and valuable contributions to management efforts. This is especially true for advances in technology in areas such as animal telemetry, global positioning systems, geographic information systems, and DNA studies of diet, reproductive success, and relatedness, to mention only a few. Closing the door on theoretical research slows or sometimes eliminates the application of essential methods for management. Such reflections seem obvious to a researcher and an academician, but I know from experience that this is not the attitude of some individuals charged with the management of natural resources. The topic of sexual segregation provides an excellent example for why some managers should reconsider long held and adverse attitudes toward basic research and consider supporting some theoretical studies.

Future Questions and Directions

Sexual segregation in ungulates has potential to bias population estimates. If the sexes are not sampled during periods of sexual aggregation, complete knowledge of their spatial distribution is needed to design protocols so that males and females are included in estimates of population size (Bowyer 2004). Population estimates and measurements of habitat use and selection often are presented for the population as a whole, with little consideration of the differing niche requirements and resulting spatial distributions of the sexes. Such an approach is shortsighted and can lead to management errors related to harvest of animals and manipulation of habitat (Bowyer 2004), but such oversights are common (Chapter 9).

A dire need exists to test hypotheses related to sexual segregation with manipulative experiments

(McCullough 1979; Kie and Bowyer 1999; Stewart et al. 2003b, 2015; Bowyer 2004; Spaeth al. 2004). Studies based largely on correlations among variables related to sexual segregation risk confusing cause and effect, and they are open to misinterpretation (Bowyer et al. 2005, 2020b; Stewart et al. 2011). This manipulative approach is critical to gaining new knowledge about why the sexes segregate, and for furthering our capability to manage and conserve ungulates. Such research, however, needs to balance the necessity for experimentation with realistic environmental conditions under which sexual segregation evolved, which accommodate the life-history requirements of free-ranging ungulates. Studies that manipulate density of populations or critical habitats have been especially instructive in gaining new insights into sexual segregation; I believe emphasis should be given to such research. Moreover, asking questions about the adaptive significance of animal behavior is essential to understanding its underpinnings—we should not rely on explanations for sexual segregation that lack such an evolutionary perspective. Darwin's theory of natural selection provides a powerful lens for viewing and understanding the natural world. Those who shun this approach ultimately may learn to appreciate the beauty and complexity of nature, but they will never unravel its intricacies.

There is a pressing need to understand how differential selection by the sexes of space, habitat, and diets (see Table 5) relates to the individual fitness consequences of those behaviors. The typical assumption is that such behaviors are adaptive, but supportive data are rare. Studies of factors related to reproductive fitness in selection of habitats similar to those of Long et al. (2016) are essential for understanding differences in sexual segregation of ungulates. This will be challenging research, but it is an important next step.

Sexual segregation of sexually dimorphic ungulates also may affect predator-prey relationships in unexpected ways (Bowyer et al. 2005). For example, the death of a male ungulate killed by a predator will have a proportionally lower influence on recruitment of young into the prey population than the death of a female (McCullough 1979; Chapter 8). Because males of dimorphic ruminants are considerably larger than females (Weckerly 1998b; Chapter 2), however, the greater amount of food larger males supply is likely to affect predator reproduction more than would smaller-bodied females or young, depending of course on the rates at which those animals are killed. Differential vulnerability of the sexes of ungulates to predators affects risk of predation and thereby the spatial distribution of the sexes and subsequent outcomes related to the kill rate of predators. Those outcomes have potential to affect predator-prey dynamics in ways that are not considered in existing models (Bowyer et al. 2005). How sexes of dimorphic ungulates are distributed spatially, and effects of this pattern on predator-prey dynamics, is a topic in urgent need of additional research.

Sexual segregation in dimorphic ungulates is near ubiquitous (Bowyer 2004, Stewart et al. 2011; Chapter 7). Sexual segregation clearly results in the partitioning of resources between males and females of the same species (see Table 5), but how interspecific competition is affected by this process is unknown. For example, how is interspecific competition among sympatric species of dimorphic ungulates influenced by smaller males of one species competing with slightly larger females of another species? Sexual differences in body size and behavior in relation to interspecific competition is not well understood. In most studies of interspecific competition, only the species is considered, with little attention given to the sex of competitors (Bowyer 2004, Stewart 2011, Bowyer et al. in press).

The manner in which the sexes segregate spatially and then aggregate for mating could affect gene flow and, consequently, the genetic diversity of populations (Bowyer 2004, Rubin and Bleich 2005). Any process that results in a noncontiguous distribution of animals on the landscape has potential to affect gene flow and genetic diversity (Chesser 1991a, b). Moreover, a common method for calculating effective population size (N_e), which is especially important

in assessing the genetic status of endangered or threatened populations, involves knowing the number of breeding males and females. Hence, N_e will be affected by accuracy of the estimated sex ratio, which might be biased by the distribution of males and females on the landscape (Rubin and Bleich 2005).

Although diseases and parasites are an unlikely cause of sexual segregation (Chapter 5), the sexes may differ in their susceptibility to disease (Rubin and Bleich 2005). Indeed, the level of disease or parasites in one sex may not be representative of the population as a whole. Moreover, movement to and from areas of sexual segregation and aggregation risks animals of one sex carrying pathogens to uninfected individuals of the other sex (Shannon et al. 2014). In addition, levels of environmental contamination may differ between sexes because some species inhabit largely separate areas for part of the year (Rubin and Bleich 2005). Most of these potentialities have not been investigated fully or considered in the framework of sexual segregation.

Anthropogenetic developments may affect patterns of sexual segregation by obstructing movements of males or females. Such changes to the landscape, including housing developments, roads, power lines, energy developments, pipelines, aqueducts, and probably numerous other changes, may influence the manner in which the sexes of ruminants use space (Bleich et al. 1997, 2016; Bowyer 2004; Rubin and Bleich 2005). Understanding requirements of the sexes should be a mandatory part of mitigation plans. There is a vital need to acquire such information for ungulates to ensure their wise management and conservation in the future.

Finally, we know that climate change poses threats to many ungulates (Chapter 1) and can result in changing distributions of animals on the landscape. How climate change relates to the manner in which the sexes partition resources, adjust to predation risk, and potentially adapt to the consequences of such behavioral and ecological changes is largely unknown. This is a topic in urgent need of research.

Relevance to Other Organisms

When I began this book, I envisioned being able to use the literature to draw inference not only about ungulates but also other organisms that sexually segregate. Table 1 illustrates the diversity of at least some species for which sexual segregation has been described. The complexity is overwhelming, and the reasons underpinning sexual segregation numerous. Rather than simply giving up on a comparative approach to understanding sexual segregation that cuts across taxa, I offer some salient reasons for not pursuing such comparisons—reasons that extend beyond the complexity of that undertaking.

Comprehending the causes of sexual segregation among organisms requires a thorough understanding of their evolutionary backgrounds, natural history, and life-history characteristics if valid comparisons are to be made across taxa. Consider the task of compiling information on the differing life histories of plants, invertebrates, and vertebrates on their propensity to sexually segregate—a challenging if not hopeless task. Moreover, it is equally important to understand the attributes of species that sexually segregate as well as those that do not. As I mentioned in Chapter 1, the literature may not reflect those patterns accurately.

Even restricting comparisons to vertebrates is problematical. There are fish that sexually segregate that do not exhibit sexual dimorphism (Griffiths et al. 2014). Sexual segregation in some birds occurs in species with reversed sexual size dimorphism (Smallwood 1987, Weimerskirch et al. 2006). Moreover, the route to sexual dimorphism for birds likely results from intersexual competition (natural selection) as opposed to sexual selection, which is typical for ungulates (Chapter 2). The requirements of an aquatic lifestyle likely place constraints on fishes that are not encountered by most birds and terrestrial mammals. Many birds face constraints with respect to flight that are not prevalent among reptiles and amphibians. Similarly, even among mammals, differences in sexual segregation among predators and herbivores likely result from dissimilarities in

life-history characteristics and evolutionary pathways. An aquatic versus a terrestrial existence must pose different obstacles that help shape sexual segregation, as do methods of locomotion, including fossorial and volant modes. Elephants and macropods have many similarities to ungulates, but overall differences in digestive morphology and physiology, and indeterminant growth of macropods, limit the usefulness of those comparisons. Both environmental and social factors effect ungulate mating systems (Bowyer et al. 2020a), which also affect the degree of sexual segregation via mechanisms such as paternal care (Chapter 8). Similar conditions likely hold for other vertebrates.

A comparative approach also is hampered by differences in definitions for sexual segregation. The manner in which sexual segregation is measured, differences in scale, and innumerable other difficulties mentioned in Chapters 3 and 4 make objective comparisons from the literature difficult. I noted previously (Bowyer 2004) that the comparative approach that has offered important strides in other disciplines of biology simply would not work for understanding sexual segregation. I am now even more convinced that this is true. Nonetheless, I offer Table 1 and my best wishes as starting points for anyone who desires to try this approach.

What about humans (*Homo sapiens*)? Aldo Leopold (1966) noted that "man brings all things to the test of himself," and I expect some readers would be disappointed if I did not at least broach the topic of sexual segregation in humans. Characteristics related to human culture and differences in size of males and females may predispose us toward sexually segregating. For instance, about 85% of human societies are polygynous (White et al. 1988) and have been so for much of our evolutionary history (Dixson 2009). Indeed, the evolution of monogamy in humans maybe a relatively recent occurrence (Plavcan 2012). Variation among human societies occurs in mating systems: slight polygyny (51.6%), general polygyny (30.6%), monogamy (16.7%), and polyandry (1.1%) (Marlowe 2000). Moreover, humans exhibit slight sexual size dimorphism—the ratio of male-to-female body weights for humans is about 1.1–1.2; a ratio that has declined over evolutionary time (Clutton-Brock 2016). Nonetheless, this ratio is outside the range for monogamous nonhuman primates (Dixson 2009). For those interested, Clutton-Brock (2016) provides a broad and useful overview of human societies.

Sexual segregation, as described for humans, involves mostly different groupings of male and female children and juveniles (Pellegrini et al. 2005). There clearly are cultural differences in associations of adults, but such patterns are unquestionably social in nature (Pellegrini et al. 2005). Differences between the sexes influence many societal practices, including legal ones (Alves 2020). A multitude of factors promote the evolution of gregariousness in humans (Clutton-Brock 2016), but such gatherings have been categorized primarily as social segregation (Pellegrini et al. 2005) and fit poorly into what we know about ungulates.

Recently, Cohen et al. (2021) argued that because women have a different baseline sensitivity to cold temperatures than men, this difference offered a basis for sexual segregation in humans as well as other species such as birds and bats (discussed in Chapter 5). No indication is provided, however, about how women and men partition space or join separate social groupings in relation to changing temperatures. Moreover, no evidence is presented for how females being more sensitive to cold than males might affect the reproductive success of the sexes. At high latitudes (40°–60°) in the Northern Hemisphere, where differences in responses to cold temperatures should be most pronounced, peak months of birth tended to be between May and September (Martinez-Bakker et al. 2014). This outcome resulted from conceptions that occurred from September to January, when temperatures were low, which runs contrary to the hypothesis that cold temperatures separated the sexes. Cold temperatures clearly did not prevent aggregation of sexes during autumn and winter. This result is concordant with my experience

from spending 18 years in interior Alaska, where social gatherings often occurred during periods of cold weather.

Given our propensity toward polygyny and slight sexual dimorphism, is there evidence of male and female adult humans spatially segregating? I reiterate that such spatial differences cannot be just the intermittent use of space that is not maintained over time (i.e., the sexes using the same space at different times). Convincing evidence that male and female humans spatially segregate is nil; however, this observation raises the question of why. Many canids also possess attributes related to sexual size dimorphism that would indicate that they should sexually segregate, but they do not (Chapter 1). I believe that the deciding factor is paternal investment in young, which prevents the strong feedbacks on intrasexual competition among males that occurs for sexually dimorphic ungulates (see Chapter 8 and Fig. 4). Female primates, including humans, provide substantial maternal care of young; human males exhibit complex patterns of paternal investment in offspring. Such behaviors include maintenance of women and their children, and various types of direct and indirect parental investment in young; such behaviors are absent from African apes (Clutton-Brock 2016). Further, humans may engage in alloparental care in which grandparents and other relatives assist in rearing young (Clutton-Brock 2016). Much of our evolutionary history and culture make spatial segregation of the sexes unlikely.

Nothing would displease me more than to have the knowledge concerning differences between the sexes, especially sexual segregation in ungulates, used as a basis or excuse for misogynistic attitudes. Although ungulates have played an important role in our evolution and culture (Chapter 1), their fascinating behaviors related to sexual segregation are not a guide for acceptable human conduct. Male investment in women and children is a unique component of human evolution and culture; it separates us from other sexually dimorphic primates and benefits men, women, and children. Women obviously are critical for the success and maintenance of our culture and provide an important foundation for our society. In no way should differences between the sexes of ungulates be used as a justification for sexism.

I initially proposed to examine patterns of sexual segregation in ungulates, its definitions, hypotheses for its occurrence, and its relevance to theory and management. I planned to clarify and organize the book into a coherent theme that would inform future research on this fascinating subject. I hope my attempt has at least accomplished some of those goals. Nevertheless, no overview of the causes and consequences of sexual segregation in ungulates is likely to be all-inclusive. My desire is that this book will lead others to become as fascinated with this subject as I am, and to continuing exploring sexual segregation where this endeavor has left off. I believe I have provided valuable contributions to understanding sexual segregation, including resolving a number of questions about this topic, but helping set directions for future research may be the more valuable contribution.

SUMMARY

The purpose of this chapter is to review factors that have hindered our understanding of sexual segregation and to suggest future directions.

1. The most fundamental difficulty with understanding the causes and consequences of sexual segregation is the lack of an operational definition of sexual segregation. The traditional definition is still the most germane for understanding sexual segregation—the differential use of space or other resources by the sexes.

2. There may be no more than 2 general explanations for sexual segregation: resources (e.g., the gastrocentric hypothesis) and risk of predation.

3. Sexual segregation has been overlooked largely because standard textbooks used in undergraduate education in the sciences, particularly ecology, management, and behavior, fail to mention sexual segregation, but a few notable exceptions exist. In addition, basic coursework in natural history often is omitted from degrees in the biological sciences.

4. I anticipate that the importance of sexual segregation in the conservation and management of animals will become more fully appreciated. Articles on this topic are finding their way into popular literature.

5. A critical need exists for more research on sexual segregation, and there should be a balance between theory and application in research—many important advances in management were discovered by theoretical studies.

6. A knowledge of sexual segregation is relevant for estimating population size of ungulates as well as their use and selection of habitat. Manipulative experiments to test hypotheses related to sexual segregation are essential.

7. More emphasis needs to be placed on how the differential selection by the sexes of space, habitat, and diets relates to the fitness consequences of those behaviors.

8. The concept of sexual segregation has yet to be addressed in studies of predator-prey dynamics, or interspecific competition.

9. Although diseases and parasites are an unlikely cause of sexual segregation, the sexes may differ in their susceptibility to diseases—a topic that is yet to be adequately addressed.

10. Human developments, such as housing, roads, power lines, energy developments, pipelines, aqueducts, and numerous other changes, may influence sexual segregation; this issue requires further study.

11. Because of a multitude of difficulties and differences in the evolution and life-history

characteristics of organisms, a comparative approach may not be viable for studying sexual segregation.

12. Humans do not spatially segregate like ungulates. Sexual segregation in humans mostly involves differences in social groupings rather than space. Sexual segregation in ungulates and differences in the behavior of the sexes should not be used as a justification for misogynistic behavior.

Literature Cited

Abouheif, E., and D. J. Fairbairn. 1997. A comparative analysis of allometry for sexual size dimorphism: assessing Rensch's rule. American Naturalist 149:540–562.

Ahmad, R., C. Mishra, N. J. Sing, R. Kaul, and Y. V. Bhatnagar. 2016. Forage security trade-offs by markhor Capra falconeri mothers. Current Science 110:1599–1564.

Ahmad, R., N. Sharma, C. Mishra, N. J. Singh, G. S. Rawat, and Y. V. Bhatnagar. 2018. Security, size or sociality: what makes markhor (Capra falconeri) sexually segregate? Journal of Mammalogy 99:55–63.

Aho, K., and R. T. Bowyer. 2015a. Confidence intervals for ratios of proportions: implications for selection ratios. Methods in Ecology and Evolution 6:121–132.

Aho, K. A., and R. T. Bowyer. 2015b. Confidence intervals for a product of proportions: application to importance values. Ecosphere 6:art230.

Airst, J., and S. Lingle. 2019. Courtship strategies of white-tailed deer and mule deer when living in sympatry. Behaviour 156:307–330.

Aivaz, A. N., and K. E. Ruckstuhl. 2011. Costs of behavioral synchrony as a potential driver behind size-assorted grouping. Behavioral Ecology 22:1353–1363.

Alados, C. L., and J. Escos. 1987. Relationships between movement rate, agonistic displacements and forage availability in Spanish ibexes (Capra pyrenaica). Biology of Behavior 12:245–255.

Albon, S. D., B. Mitchel, and B. W. Stains. 1983. Fertility and body weight in female red deer. Journal of Animal Ecology 52:969–980.

Alderman, J. A., P. R. Krausman, and B. D. Leopold. 1989. Diel activity of female desert bighorn sheep in western Arizona. Journal of Wildlife Management 53:264–271.

Alexander, R. D. 1974. The evolution of social behavior. Annual Review of Ecology and Systematics 5:325–383.

Alkali, H. A., B. F. Muhammad, A. A. Njidda, M. Abubaker, and M. I. Ghude. 2017. Relative forage preference by camel (Camelus dromedarius) as influenced by season, sex and age in the Sahel zone of north western Nigeria. African Journal of Agricultural Research 12:1–5.

Allen, D. L. 1954. Our wildlife legacy. Funk and Wagnalls Company, New York, New York, USA.

Alonso, J. C., I. Salgado, and C. Palacín. 2016. Thermal tolerance may cause sexual segregation in sexually dimorphic species living in hot environments. Behavioral Ecology 27:717–724.

Alonso-Fernádez, A., J. Otero, R. Banón, J. M. Campelos, J. Santos, and G. Mucientes. 2017. Sex ratio variation in an exploited population of common octopus: ontogenic shifts and spatio-temporal dynamics. Hydrobiologia 794:1–16.

Altmann, J. 1974. Observation study of behavior: sampling methods. Behaviour 49:227–267.

Altringham, J. D., and P. Senior. 2005. Social systems and the ecology of bats. Pages 280–302 in K. E. Ruckstuhl and P. Neuhaus, editors. Sexual segregation in vertebrates: ecology of the two sexes. Cambridge University Press, Cambridge, United Kingdom.

Alves, J., A. A. da Silva, A. M. V. M. Soares, and C. Fonseca. 2013. Sexual segregation in red deer: is social behaviour more important than habitat preferences? Animal Behaviour 85:501–509.

Alves, J., A. A. da Silva, A. M. V. M. Soares, and C. Fonseca. 2014. Spatial and temporal habitat use and selection by red deer: the use of direct and indirect methods. Mammalian Biology 79:338–342.

Alves, M. G. 2020. As leis e a segregação: como as leis influenciaram a segregação sexual na sociedade patriarcal contemporânea (Laws and segregation: how laws have influenced sexual segregation in contemporary patriarchal society). Brazilian Journal of Development 6:8019–8028.

Alvis, Á., and J. Pérez-Torres. 2020. A difference between sexes: temporal variation in the diet of *Carollia perspicillata* (Chiroptera, Phyllostomidae) at the Macaregua cave, Santander (Colombia). Animal Biodiversity and Conservation 43:27–35.

Andelt, W. F., J. G. Kie, F. F. Knowlton, and K. Cardwell. 1987. Variation in coyote diets associated with season and successional changes in vegetation. Journal of Wildlife Management 51:273–277.

Andersson, M. 1994. Sexual selection. Princeton University Press, Princeton, New Jersey, USA.

Angell, R. L., R. K. Butlin, and J. D. Altringham. 2013. Sexual segregation and flexible mating patterns in temperate bats. PLoS ONE 8:e54194.

Apollonio, M., F. Brivio, I. Rossi, B. Bassano, and S. Grignolio. 2013. Consequences of snowy winters on male mating strategies and reproduction in a mountain ungulate. Behavioural Processes 98:44–50.

Apollonio, M., S. Ciuti, and S. Luccarini. 2005. Long-term influence of human presence on spatial sexual segregation in fallow deer (*Dama dama*). Journal of Mammalogy 86:937–946.

Apollonio, M., M. Festa-Bianchet, and F. Mar. 1989. Correlates of copulatory success in a fallow deer lek. Behavioral Ecology and Sociobiology 25:89–97.

Apollonio, M., E. Merli, R. Chirichella, B. Pokorny, A. Alagić, K. Flajšman, and P. A. Stephens. 2020. Capital-income breeding in male ungulates: causes and consequences of strategy differences among species. Frontiers in Ecology and Evolution 8:521767.

Arkhipkin, A., and D. Middleton. 2002. Sexual segregation in ontogenetic migrations by the squid *Loligo gahi* around the Falkland Islands. Bulletin of Marine Science 71:109–127.

Arnold, S. J. 1994. Bateman's principles and the measurement of sexual selection in plants and animals. American Naturalist 144:S126–S149.

Arnold, T. W. 2010. Uninformative parameters and model selection using Akaike's information criterion. Journal of Wildlife Management 74:1175–1178.

Atkins, J. L., R. A. Long, J. Pansu, J. H. Daskin, A. B. Potter, M. E. Stalmans, C. E. Tarnita, and R. M. Pringle. 2019. Cascading impacts of large-carnivore extirpation in an African ecosystem. Science 364:173–177.

Atwood, T. C., E. M. Gese, and K. E. Kunkel. 2009. Spatial partitioning of predation risk in a multiple predator–multiple prey system. Journal of Wildlife Management 73:876–884.

Austin, R. E., F. De Pascalis, S. C. Votier, J. Haakonsson, J. P. Y. Arnold, G. Ebanks-Petrie, J. Newton, and J. A. Green. 2021. Interspecific and intraspecific foraging differentiation of neighbouring tropical seabirds. Movement Ecology 9:27.

Avgar, T., G. Street, and J. M. Fryxell. 2014. On the adaptive benefits of mammal migration. Canadian Journal of Zoology 92:481–490.

Azorit, C., S. Tellado, A. Oya, and J. Moro. 2012. Seasonal and specific diet variations in sympatric red and fallow deer of southern Spain: a preliminary approach to feeding behaviour. Animal Production Science 52:720–727.

Baird R. W., M. B. Hanson, and L. M. Dill. 2005. Factors influencing the diving behaviour of fish-eating killer whales: sex differences and diel and interannual variation in diving rates. Canadian Journal of Zoology 83:257–267.

Bajzak, C. E., S. D. Côté, M. O. Hammill, and G. Stenson. 2009. Intersexual differences in the postbreeding foraging behaviour of the Northwest Atlantic hooded seal. Marine Ecology Progress Series 385:285–294.

Bañuelos, M.-J., M. Quevedo, and J.-R. Obeso. 2008. Habitat partitioning in endangered Cantabrian capercaillie *Tetrao urogallus cantabricus*. Journal of Ornithology 149:254–252.

Barash, D. P. 1982. Sociobiology and behavior. Second edition. Elsevier Scientific, New York, New York, USA.

Barboza, P. S., and R. T. Bowyer. 2000. Sexual segregation in dimorphic deer: a new gastrocentric hypothesis. Journal of Mammalogy 81:473–489.

Barboza, P. S., and R. T. Bowyer. 2001. Seasonality of sexual segregation in dimorphic deer: extending the gastrocentric model. Alces 37:275–292.

Barboza, P. S., D. W. Hartbauer, W. E. Hauer, and J. E. Blake. 2004. Polygynous mating impairs body condition and homeostasis in male reindeer (*Rangifer tarandus tarandus*). Journal of Comparative Physiology B 174:309–317.

Barboza, P. S., K. L. Parker, and I. D. Hume. 2009. Integrative wildlife nutrition. Springer-Verlag, Berlin, Germany.

Barboza, P. S., R. D. Shively, D. D. Gustine, and J. A. Addison. 2020. Winter is coming: conserving body protein in female reindeer, caribou, and muskoxen. Frontiers in Ecology and Evolution 8:150.

Barboza, P. S., L. L. Van Someren, D. D. Gustine, and M. S. Bret-Harte. 2018. The nitrogen window for arctic herbivores: plant phenology and protein gain of migratory caribou (*Rangifer tarandus*). Ecosphere 9:e02073.

Barbraud, C., K. Delord, A. Kato, P. Bustamante, and Y. Cherel. 2019. Sexual segregation in a highly pagophilic and sexually dimorphic marine predator. Peer Community in Ecology. https://doi.org/10.1101/472431.

Bårdsen, B.-J., and T. Tveraa. 2012. Density-dependence vs. density independence—linking reproductive allocation to population abundance and vegetation greenness. Journal of Animal Ecology 81:364–376.

Barnett, A., K. G. Abrantes, J. D. Stevens, and J. M. Semmens. 2011. Site fidelity and sex-specific migration in a mobile apex predator: implication for conservation and ecosystem dynamics. Animal Behaviour 81:1039–1048.

Barrett, S. C. H., and J. Hough. 2013. Sexual dimorphism in flowering plants. Journal of Experimental Biology 64:67–82.

Barría, E. M., R. D. Sepúlveda, and C. G. Jara. 2011. Morphological variation in *Aegla* leach (Decapoda: Reptantia: Aeglidae) from central-southern Chile: interspecific differences, sexual dimorphism, and spatial segregation. Journal of Crustacean Biology 31:231–239.

Barrionuevo, M., J. Ciacio, A. Steinfurth, and E. Frere. 2020. Geolocation and stable isotopes indicate habitat segregation between sexes in Magellanic penguins during the winter. Journal of Avian Biology 51:e02325.

Bartelt, P. E., C. R. Peterson, and R. W. Klaver. 2004. Sexual differences in the post-breeding movements and habitats selected by western toads (*Bufo boreas*) in southeastern Idaho. Herpetologica 60:455–467.

Barten, N. L., R. T. Bowyer, and K. J. Jenkens. 2001. Habitat use by female caribou: tradeoffs associated with parturition. Journal of Wildlife Management 65:77–92.

Bartlam-Brooks, H., M. Bonyongo, and S. Harris. 2011. Will reconnecting ecosystems allow long-distance mammal migrations to resume? A case study of a zebra *Equus burchelli* migration in Botswana. Oryx 45:210–216.

Bartmann, R. M., G. C. White, and L. H. Carpenter. 1992. Compensatory mortality in a Colorado mule deer population. Wildlife Monographs 121:1–39.

Baskin, L., and K. Danell. 2003. Ecology of ungulates: a handbook of species in eastern Europe and northern and central Asia. Springer-Verlag, Berlin, Germany.

Bateman, A. J. 1948. Intra-sexual selection in *Drosophila*. Heredity 2:349–368.

Baylis, A. M. M., R. A. Orben, D. P. Costa, J. P. Y. Arnould, and I. J. Staniland. 2016. Sexual segregation in habitat use is smaller than expected in a highly dimorphic marine predator, the southern sea lion. Marine Ecology Progress Series 554:201–211.

Beck, C. A., W. D. Bowen, J. I. McMillan, and S. J. Iverson. 2003. Sex differences in diving at multiple temporal scales in a size-dimorphic capital breeder. Journal of Animal Ecology 72:979–993.

Beck, C. A., S. J. Iverson, W. D. Bowen, and W. Blanchard. 2007. Sex differences in grey seal diet reflect seasonal variation in foraging behaviour and reproductive expenditure: evidence from quantitative fatty acid signature analysis. Journal of Animal Ecology 76:490–502.

Beck, J. L., J. M. Peek, and E. K. Strand. 2006. Estimates of elk summer range nutritional carrying capacity constrained by probabilities of habitat selection. Journal of Wildlife Management 70:283–294.

Beerman, A., E. Ashe, K. Preedy, and R. Williams. 2015. Sexual segregation when forging in an extremely social killer whale population. Behavioral Ecology and Sociobiology 70:189–198.

Beier, P. 1987. Sex differences in quality of white-tailed deer diets. Journal of Mammalogy 68: 323–329.

Beier, P., and D. R. McCullough. 1990. Factors influencing white-tailed deer activity patterns and habit use. Wildlife Monographs 108:1–51.

Bell, R. H. V. 1970. The use of the herb layer by grazing ungulates in the Serengeti. Pages 111–113 *in* A. Watson, editor. Animal populations in relation to their food resources. Blackwell, Oxford, United Kingdom.

Belovsky, G. E. 1981. Diet optimization in a generalist herbivore: the moose. Theoretical Population Biology 14:105–134.

Ben-David, M., and E. A. Flaherty. 2012. Stable isotopes in mammalian research: a beginner's guide. Journal of Mammalogy 93:312–328.

Ben-David, M., E. Shochat, and L. G. Adams. 2001. The utility of stable isotope analysis in studying the foraging ecology of herbivores: examples from moose and caribou. Alces 37:421–434.

Bennett, R. E., A. D. Rodewald, and K. V. Rosenberg. 2019. Overlooked sexual segregation of habitat exposes female migratory landbirds to threats. Biological Conservation 240:art 108226.

Benoist, S., M. Garel, J.-M. Cugansse, and P. Planchard. 2013. Human disturbances, habitat characteristics and social environment generate sex-specific responses in vigilance of Mediterranean mouflon. PLoS ONE 8:e82960.

Berger, J. 1978. Group size, foraging, and antipredator ploys: an analysis of bighorn sheep decisions. Behavioral Ecology and Sociobiology 4:91–99.

Berger, J. 1986. Wild horses of the Great Basin: social competition and population size. University of Chicago Press, Chicago, Illinois, USA.

Berger, J. 1991. Pregnancy incentives, predation constraints and habitat shifts: experimental and field evidence for wild bighorn sheep. Animal Behaviour 41:61–77.

Berger, J. 2004. The last mile: how to sustain long-distance migrations in mammals. Conservation Biology 18:320–331.

Berger, J. 2010. Fear-mediated food webs. Pages 275–253 *in* J. Terborgh and J. A. Estes, editors. Trophic cascades: predators, prey, and the changing dynamics of nature. Island Press, Washington, DC, USA.

Berger, J. 2012. Estimation of body-size traits by photogrammetry in large mammals to inform conservation. Conservation Biology 26:769–777.

Berger, J. 2018. Extreme conservation: life at the edges of the world. University of Chicago Press, Chicago, Illinois, USA.

Berger, J., and C. Cunningham. 1994. Bison: conservation in small populations. Columbia University Press, New York, New York, USA.

Berger, J., and M. E. Gompper. 1999. Sex ratios in extant ungulates: products of contemporary predation or past life histories? Journal of Mammalogy 80:1084–1113.

Berger, J., E. Cheng, A. Kang, M. Krebs, L. Li, Z. X. Lu, and G. B. Schaller. 2014. Sex differences in ecology of wild yaks at high elevation in the Kekexili Reserve, Tibetan Qinghai Plateau, China. Journal of Mammalogy 95:638–645.

Berger, J., S. Dulamtseren, S. Cain, D. Enkkhbileg, P. Lichtman, Z. Namshir, G. Wingard, and R. Reading. 2001a. Back-casting sociality in extinct species: new perspectives using mass death assemblages and sex ratios. Proceedings of the Royal Society of London, B 268:131–139.

Berger, J., C. Hartway, A. Gruzdev, and M. Johnson. 2018. Climate degradation and extreme icing events constrain life in cold-adapted mammals. Scientific Reports 8:1156.

Berger, J., P. B. Stacy, L. Bellis, and M. P. Johnson. 2001b. A mammalian predator-prey imbalance: grizzly bear and wolf extinction affect avian neotropical migrants. Ecological Applications 11:947–960.

Berger, J., J. E. Swenson, and I.-L. Persson. 2001c. Recolonizing carnivores and naïve prey: conservation lessons from Pleistocene extinctions. Science 291:1036–1039.

Berger, J., T. Wangchuck, C. Bricento, A. Vilia, and J. W. Lambert. 2020. Disassembled food webs and messy projections: modern ungulate communities in the face of human population growth. Frontiers in Ecology and Evolution 8:128.

Bergeron, P., S. Grignolio, M. Apollonio, B. Shipley, and M. Festa-Bianchet. 2010. Secondary sexual characters signal fighting ability and determine social rank in alpine ibex (*Capra ibex*). Behavioral Ecology and Sociobiology 64:1299–1307.

Bergerud, A. T., and J. P. Elliot. 1986. Dynamics of caribou and wolves in northern British Columbia. Canadian Journal of Zoology 64:1515–1529.

Bergerud, A. T., and R. E. Page. 1987. Displacement and dispersion of parturient caribou at calving as antipredator tactics. Canadian Journal of Zoology 65:1597–1606.

Bergerud, A. T., H. E. Butler, and D. R. Miller. 1984. Antipredator tactics of calving caribou: dispersion in mountains. Canadian Journal of Zoology 62:1566–1575.

Bergman, E. J., P. F. Doherty Jr., G. C. White, and A. A. Holland. 2015. Density dependence in mule deer: a review of evidence. Wildlife Biology 21:18–29.

Berini, J. L., and C. Badgley. 2017. Diet segregation in American bison (*Bison bison*) of Yellowstone National Park (Wyoming, USA). BMC Ecology 17:27.

Berteaux, D. 1993. Female-biased mortality in a sexually dimorphic ungulate: feral cattle of Amsterdam Island. Journal of Mammalogy 74:732–737.

Bertiller, M. B., C. L. Sain, A. J. Bisigato, F. R. Coronato, O. J. Ares, and P. Graff. 2002. Spatial sex segregation in the dioecious grass *Poa ligularis* in northern Patagonia: the role of environmental patchiness. Biodiversity Conservation 11:69–84.

Bertram, B. C. R. 1973. Lion population regulation. East African Wildlife Journal 11:215–225.

Bierzychudek, P., and V. Eckhart. 1988. Spatial segregation of the sexes of dioecious plants. American Naturalist 132:34–43.

Biggerstaff, M. T., M. A. Lashley, M. C. Chitwood, C. E. Moorman, and C. S. DePerno. 2017. Sexual segregation of forage patch use: support for the social-factors and predation hypotheses. Behavioural Processes 136:36–42.

Bishop, C. J., G. C. White, D. J. Freddy, B. E. Watkins, and T. R. Stephenson. 2009. Effect of enhanced nutrition on mule deer population rate of change. Wildlife Monographs 172:1–28.

Bissonette, J. A. 1982. Social behavior and ecology of the collared peccary in Big Bend National Park. National Park Service Monograph 16:1–85.

Bizzarro, J. J., K. M. Broms, M. G. Logsdon, D. A. Ebert, M. M. Yoklavich, L. A. Kuhnz, and A. P. Summer. 2014. Spatial segregation in eastern North Pacific skate assemblages. PLoS ONE 9:e109907.

Bjøneraass, K., I. Herfindal, E. J. Solberg, B.-E. Sæther, B. van Moorter, and C. M. Rolandsen. 2012. Habitat quality influences population distribution, individual space use and functional responses in habit selection by a large herbivore. Oecologia 168:231–243.

Blanckenhorn, W. U. 2005. Behavioral causes and consequences of sexual size dimorphism. Ethology 111:977–1016.

Blank, D. A., K. E. Ruckstuhl, and W. Yang. 2012. Sexual segregation in goitered gazelles (*Gazella subgutturosa*). Canadian Journal of Zoology 90:955–960.

Bleich, V. C. 1999. Mountain sheep and coyotes: patterns of predator evasion in a mountain ungulate. Journal of Mammalogy 80:283–289.

Bleich, V.C., and M. W. Oehler Sr. 2007. Wildlife education in the United States: thoughts from agency biologists. Wildlife Society Bulletin 28:542–545.

Bleich, V. C., and B. M. Pierce. 2001. Accidental mass mortality of migrating mule deer. Western North American Naturalist 61:124–125.

Bleich, V. C., R. T. Bowyer, and J. D. Wehausen. 1997. Sexual segregation in mountain sheep: resources or predation? Wildlife Monographs 134:1–50.

Bleich, V. C., G. A. Sargeant, and B. P. Wiedmann. 2018. Ecotypic variation in population dynamics of reintroduced bighorn sheep: implications for management. Journal of Wildlife Management 82:8–18.

Bleich, V. C., J. C. Whiting, J. G. Kie, and R. T. Bowyer. 2016. Roads, routes, and rams: does sexual segregation contribute to anthropogenic risk in a desert-dwelling ungulate? Wildlife Research 43:380–388.

Bliss, L. M., and F. L Weckerly. 2016. Habitat use by male and female Roosevelt elk in northwestern California. California Fish and Game 102:8–16.

Blouin, J., J. Debow, E. Rosenblatt, C. Alexander, K. Gieder, N. Fortin, J. Murdoch, and T. Donovan. 2021. Modeling moose habitat use by age, sex, and season in Vermont, USA using high-resolution lidar and national land cover data. Alces 57:71–98.

Blundell, G. M., M. Ben-David, P. Groves, R. T. Bowyer, and E. Geffen. 2004. Kinship and sociality in coastal river otters: are the related? Behavioral Ecology 15:705–714.

Bocci, A., I. Angelini, P. Brambilla, A. Monaco, and S. Lovari. 2012. Shifter and resident red deer: intrapopulation and intersexual behavioural diversities in a predator-free area. Wildlife Research 39:573–582.

Bon, R., and R. Campan. 1989. Social tendencies of the Corsican mouflon (Ovis ammon musimon) in the Caroux-Espinouse Massif (south of France). Behavioural Processes 19:57–78.

Bon, R., and R. Campan. 1996. Unexplained sexual segregation in polygamous ungulates: a defense of an ontogenetic approach. Behavioural Processes 38:131–154.

Bon, R., J.-L. Deneubourg, J.-F. Gerad, and P. Bichelena. 2005. Sexual segregation in ungulates: from individual mechanisms to collective patterns. Pages 180–199 in K. E. Ruckstuhl and P. Neuhaus, editors. Sexual segregation in vertebrates: ecology of the two sexes. Cambridge University Press, Cambridge, United Kingdom.

Bon, R., C. Rideau, J. C. Villaret, and J. Joachim. 2001. Segregation is not only a matter of sex in alpine ibex (Capra ibex ibex). Animal Behaviour 62:495–504.

Bonenfant, C., J.-M. Gaillard, T. Coulson, M. Festa-Bianchet, A. Loison, M. G. Leif, E. Loe, P. Blanchard, N. Pettorelli, N. Owen-Smith, et al. 2009. Empirical evidence of density-dependence in populations of large herbivores. Advances in Ecological Research 41:313–357.

Bonenfant, C., J.-M. Gaillard, S. Dray, A. Loison, and M. Royer. 2007. Testing sexual segregation and aggregation: old ways are best. Ecology 88:3202–3208.

Bonenfant, C., L. E. Loe, A. Mysterud, R. Langvtatn, N. C. Stenseth, J.-M. Gaillard, and F. Klein. 2004. Multiple causes of sexual segregation in European red deer: enlightenments from varying breeding phenology at high and low latitude. Proceedings or the Royal Society of London B 271:883–892.

Boner, C. 1861. Forest creatures. Longman, Green, Longman, and Roberts, London, United Kingdom (original not seen, cited in Darwin 1871).

Bourgoin, G., P. Marchand, A. J. M. Hewiston, K. E. Ruckstuhl, and M. Garel. 2018. Social behaviour as a predominant driver of sexual age-dependent and reproductive segregation in Mediterranean mouflon. Animal Behaviour 136:87–100.

Bowden, D. C., A. E. Anderson, and D. E. Medin. 1984. Sampling plans for mule deer sex and age ratios. Journal of Wildlife Management 48:500–509.

Bowers, M. A., and H. D. Smith. 1979. Differential habitat utilization by sexes of the deer mouse, Peromyscus municulatus. Ecology 60:869–875.

Bowyer, J. W., and R. T. Bowyer. 1997. Effects of previous browsing on the selection of willow stems by Alaskan moose. Alces 33:11–18.

Bowyer, R. T. 1981. Activity, movement, and distribution of Roosevelt elk during rut. Journal of Mammalogy 62:574–582.

Bowyer, R. T. 1984. Sexual segregation in southern mule deer. Journal of Mammalogy 65:410–417.

Bowyer, R. T. 1986a. Antler characteristics as related to social status of male southern mule deer. Southwestern Naturalist 31:289–298.

Bowyer, R. T. 1986b. Habitat selection by southern mule deer. California Fish and Game 72:153–169.

Bowyer, R. T. 1987. Coyote group size relative to predation on mule deer. Mammalia 51:515–526.

Bowyer, R. T. 1991. Timing of parturition and lactation in southern mule deer. Journal of Mammalogy 72:138–145.

Bowyer, R. T. 2004. Sexual segregation in ruminants: definitions, hypotheses and implications for conservation and management. Journal of Mammalogy 85:1039–1052.

Bowyer, R. T., and V. C. Bleich. 1984. Effects of cattle grazing on selected habitats of southern mule deer. California Fish and Game 70:240–247.

Bowyer, R. T., and J. G. Kie. 2004. Effects of foraging activity on sexual segregation in mule deer. Journal of Mammalogy 85:498–504.

Bowyer, R. T., and J. G. Kie. 2006. Effects of scale on interpreting life-history characteristics of ungulates and carnivores. Diversity and Distributions 12:244–257.

Bowyer, R. T., and J. G. Kie. 2009. Thermal landscapes and resource selection by black-tailed deer: implications for large herbivores. California Fish and Game 95:128–139.

Bowyer, R. T., and D. W. Kitchen. 1987. Sex and age class differences in vocalization of Roosevelt elk during rut. American Midland Naturalist 118:225–235.

Bowyer, R. T., and J. A. Neville. 2003. Effects of browsing history by Alaskan moose on regrowth and quality of feltleaf willow. Alces 39:193–202.

Bowyer, R. T., V. C. Bleich, P. R. Krausman, and J.-M. Gaillard. 2021. Editorial: advances in ungulate ecology. Frontiers in Ecology and Evolution 9:675265.

Bowyer, R. T., V. C. Bleich, X. Manteca, J. C. Whiting, and K. M. Stewart. 2007. Sociality, mate choice, and timing of mating in American bison (Bison bison): effects of large males. Ethology 113:1048–1060.

Bowyer, R. T., V. C. Bleich, K. M. Stewart, J. C. Whiting, and K. L. Monteith. 2014. Density dependence in ungulates: a review of causes and consequences, with some clarifications. California Fish and Game 100:550–572.

Bowyer, R. T., M. S. Boyce, J. R. Goheen, and J. L. Rachlow. 2019. Conservation of the world's mammals: status, protected areas, community efforts, and hunting. Journal of Mammalogy 100:923–941.

Bowyer, R. T., J. G. Kie, D. K. Person, and K. L. Monteith. 2013. Metrics of predation: perils of predator-prey ratios. Acta Theologica 58:329–340.

Bowyer, R. T., J. G. Kie, and V. Van Ballenberghe. 1996. Sexual segregation in black-tailed deer: effects of scale. Journal of Wildlife Management 60:10–17.

Bowyer, R. T., J. G. Kie, and V. Van Ballenberghe. 1998a. Habitat selection by neonatal black-tailed deer: climate, forage, or risk of predation? Journal of Mammalogy 79:415–425.

Bowyer, R. T., D. R. McCullough, and G. E. Belovsky. 2001a. Causes and consequences of sociality in mule deer. Alces 37:371–402.

Bowyer, R. T., D. R. McCullough, J. L. Rachlow, S. Ciuti, and J. C. Whiting. 2020a. Evolution of ungulate mating systems: integrating social and environmental factors. Ecology and Evolution 10:5160–5178.

Bowyer, R. T., M. C. Nicholson, E. M. Molvar, and J. B. Faro. 1999a. Moose on Kalgin Island: are density-dependent processes related to harvest? Alces 35:73–89.

Bowyer, R. T., D. K. Person, and B. M. Pierce. 2005. Detecting top-down versus bottom-up regulation of ungulates by large carnivores: implication for conservation of biodiversity. Pages 342–361 in J. C. Ray, K. H. Redford, R. S. Steneck, and J. Berger, editors. Large carnivores and the conservation of biodiversity. Island Press, Washington, DC, USA.

Bowyer, R. T., B. M. Pierce, L. K. Duffy, and D. A. Haggstrom. 2001b. Sexual segregation in moose: effects of habitat manipulation. Alces 37:109–122.

Bowyer, R. T., J. L. Rachlow., K. M. Stewart, and V. Van Ballenberghe. 2011. Vocalizations by Alaskan moose: female incitation of male aggression. Behavioral Ecology and Sociobiology 65: 2251–2260.

Bowyer, R. T., J. L. Rachlow, V. Van Ballenberghe, and R. D. Guthrie. 1991. Evolution of a rump patch in Alaskan moose: an hypothesis. Alces 27:12–23.

Bowyer, R. T., K. M. Stewart, V. C. Bleich, J. C. Whiting, K. L. Monteith, M. E. Blum, and T. N. LaSharr. 2020b. Metrics of harvest for ungulate populations: misconceptions, lurking variables, and prudent management. Alces 56:15–38.

Bowyer, R. T., K. M. Stewart, J. W. Cain III, and B. R. McMillan. In press. Competition with other ungulates. In J. R. Heffelfinger and P. R. Krausman, editors. Ecology and management of mule deer and black-tailed deer in North America. CRC Press, Boca Raton, Florida, USA.

Bowyer, R. T., K. M. Stewart, B. M. Pierce, K. J. Hundertmark, and W. C. Gasaway. 2002a. Geographical variation in antler morphology of Alaskan moose: putative effects of habitat and genetics. Alces 38:155–165.

Bowyer, R. T., K. M. Stewart, S. A. Wolfe, G. M. Blundell, K. L. Lehmkuhl, P. J. Joy, T. J. McDonough, and J. G. Kie. 2002b. Assessing sexual segregation in deer. Journal of Wildlife Management 66:536–544.

Bowyer, R. T., V. Van Ballenberghe, and J. G. Kie. 1997. The role of moose in landscape processes: effects of biogeography, population dynamics, and predation. Pages 265–287 in J. A. Bissonette, editor. Wildlife and landscape ecology: effects of pattern and scale. Springer-Verlag, New York, New York, USA.

Bowyer, R. T., V. Van Ballenberghe, and J. G. Kie. 1998b. Timing and synchrony of parturition in Alaskan moose: long-term versus proximal effects of climate. Journal of Mammalogy 79:1332–1344.

Bowyer, R. T., V. Van Ballenberghe, J. G. Kie, and J. A. K. Maier. 1999b. Birth-site selection by Alaskan moose: maternal strategies for coping with a risky environment. Journal of Mammalogy 80:1070–1083.

Boyce, M. S. 1988. Evolution of life histories: theory and patterns from mammals. Pages 3–30 in M. S. Boyce, editor. Evolution of life histories of mammals: theory and pattern. Yale University Press, New Haven, Connecticut, USA.

Boyce, M. S. 1989. The Jackson elk herd: intensive wildlife management in North America. Cambridge University Press, Cambridge, United Kingdom.

Boyce, M. S. 2006. Scale for resource selection functions. Diversity and Distributions 12:269–276.

Boyce M. S., A. R. E. Sinclair, and G. C. White. 1999. Seasonal compensation of predation and harvesting. Oikos 87: 419–426.

Braccini, M., and S. Taylor. 2016. The spatial segregation of sharks from western Australia. Royal Society Open Science 3:160306.

Brashares, J. S., P. Arcese, and M. K. Sam. 2001. Human demography and reserve size predict wildlife extinction in West Africa. Proceedings of the Royal Society of London B 268:2473–2478.

Brashares, J. S., T. Garland, and P. Arcese. 2000. Phylogenetic analysis of coadaptation in behavior, diet, and body size in the African antelope. Behavioral Ecology 11:452–463.

Bravo, C., C. Ponce, L. M. Bautista, and J. C. Alonso. 2016. Dietary divergence in the most sexually size-dimorphic bird. Auk 133:178–197.

Breiman, L. 2001. Random forests. Machine Learning 45:5–32.

Brewer, C. E., and L. A. Harveson. 2007. Diets of bighorn sheep in the Chihuahuan Desert, Texas. Southwestern Naturalist 52:97–103.

Brinkman T. J., and K. J. Hundertmark. 2009. Sex identification of northern ungulates using low quality and quantity DNA. Conservation Genetics 10.1189–1193.

Brockmann, S. F., and D. H. Pletscher. 1993. Winter segregation by the sexes of white-tailed deer. Western Journal of Forestry 8:28–33.

Bro-Jørgensen, J. 2002. Overt female mate competition and preference for central males in a lekking antelope. Proceedings of the National Academy of Sciences of the United States of America 99:9290–9293.

Bro-Jørgensen, J. 2007. The intensity of sexual selection predicts weapon size in male bovids. Evolution 61:1316–1326.

Brotherton, P. N. M., and M. B. Manser. 1997. Female dispersion and the evolution of monogamy in the dik-dik. Animal Behaviour 54:1413–1424.

Brown, M. B., and D. T. Bolger. 2020. Male-biased partial migration in a giraffe population. Frontiers in Ecology and Evolution 7:524.

Brown, J. S., and B. P. Kolter. 2004. Hazardous duty pay and the foraging cost of predation. Ecology Letters 7:999–1014.

Bryant, J. P., F. S. Chapin, and D. R. Klein. 1983. Carbon/nutrient balance of boreal plants in relation to herbivory. Oikos 40:357–368.

Bultć, G., M.-A. Gravel, and G. Blouin-Demers. 2008. Intersexual niche divergence in northern map turtles (Graptemys geographica): the roles of diet and habitat. Canadian Journal of Zoology 86:1235–1243.

Bunn, H. T., and A. N. Gurtov. 2014. Prey mortality profiles indicate that early Pleistocene Homo at Olduvai was an ambush predator. Quaternary International 322–323:44–53.

Bunnell, F. L., and M. P. Gillingham. 1985. Foraging behavior: dynamics of dining out. Pages 53–79 in R. J. Hudson and R. G. White, editors. Bioenergetics of wild herbivores. CRC Press, Boca Raton, Florida, USA.

Burnham, K. P., and D. R. Anderson. 2002. Model selection and multimodel inference: a practical information-theoretic approach. Second edition. Springer Science, New York, New York, USA.

Burns, J. G., and R. C. Ydenberg. 2002. The effects of wing loading and gender on the escape flights of least sandpipers (Calidris minutilla) and western sandpipers (Calidris mauri). Behavioral Ecology and Sociobiology 52:128–136.

Burns, L. E., and H. G. Broders. 2015. Maximizing mating opportunities: higher autumn swarming activity in male versus female Myotis bats. Journal of Mammalogy 96:1326–1336.

Burt, W. H. 1943. Territoriality and home range concepts as applied to mammals. Journal of Mammalogy 24:346–352.

Byers, J. A. 1997. American pronghorn: social adaptations and the ghosts of predators past. University of Chicago Press, Chicago, Illinois, USA.

Byers, J. A., and D. W. Kitchen. 1988. Mating system shift in a pronghorn population. Behavioral Ecology and Sociobiology 22: 355–360.

Calado, J. G., V. H. Paiva, F. R. Ceia, P. Gomes, J. A. Ramos, and A. Velando. 2020. Stable isotopes reveal rear-round sexual trophic segregation in four yellow-legged gull colonies. Marine Biology 167:art.65.

Calenge, C. 2006. The package "adehabitat" for the R software: a tool for the analysis of space and habitat use by animals. Ecological Modelling 197:516–519.

Calhim, S., J. Shi, and R. I. M. Dunbar. 2006. Sexual segregation among feral goats: testing between alternative hypothesis. Animal Behaviour 72:31–41.

Cameron, G. N., and S. R. Spencer. 2008. Mechanisms of habitat selection by the hispid cotton rat (Sigmodon hispidus). Journal of Mammalogy 89:126–131.

Cameron, R. D., and K. R. Whitten. 1979. Seasonal movements and sexual segregation of caribou determined by aerial survey. Journal of Wildlife Management 43:626–633.

Cameron, R. D., W. T. Smith, R. G. White, and B. Griffith. 2005. Central Arctic caribou and petroleum development: distributional, nutritional, and reproductive implications. Arctic 58:1–9.

Camp, M. J., J. L. Rachlow, B. A. Woods, T. R. Johnson, and L. A. Shipley. 2012. When to run and when to hide: the influence of concealment, visibility, and proximity to refugia on perceptions of risk. Ethology 118:1010–1017.

Carbyn, L. N., S. M. Oosenbrug, and D. W. Anions. 1993. Wolves, bison . . . and the dynamics related to the Peace-Athabasca Delta in Canada's Wood Buffalo National Park. Circumpolar Research Series 4, University of Alberta, Edmonton, Alberta, Canada.

Cardillo, M., G. M. Mace, K. E. Jones, J. Bielby, O. R. Bininda-Emonds, W. Sechrest, C. D. Orme, and A. Purvis. 2005. Multiple causes of high extinction risk in large mammal species. Science 309:1239–1241.

Caro, T. M. 1986. The functions of stotting: a review of the hypotheses. Animal Behaviour 34: 649–662.

Caro, T. M. 1994. Cheetahs of the Serengeti Plains: group living in an asocial species. University of Chicago Press, Chicago, Illinois, USA.

Caro, T. M. 2005. Antipredator defenses in birds and mammals. University of Chicago Press, Chicago, Illinois, USA.

Caro, T. M., C. M. Graham, C. J. Stoner, and M. M. Flores. 2003. Correlates of horn and antler shape in bovids and cervids. Behavioral Ecology and Sociobiology 55:32–41.

Carranza, J., F. Alvarez, and T. Redondo. 1990. Territoriality as a mating strategy in red deer. Animal Behaviour 40:79–88.

Cassini, M. H. 2017. Role of fecundity selection on the evolution of sexual size dimorphism in mammals. Animal Behaviour 128:1–4.

Cassini, M. H. 2020a. A mixed model of the evolution of polygyny and sexual size dimorphism in mammals. Mammal Review 50:112–120.

Cassini, M. H. 2020b. Sexual size dimorphism and sexual selection in artiodactyls. Behavioral Ecology 31:792–797.

Cassini, M. H. 2022. Evolution of size dimorphism and sexual segregation in artiodactyls: the chicken or the egg? Mammalian Biology 102:131–141.

Cassirer, F. E., and A. R. E. Sinclair. 2007. Dynamics of pneumonia in a bighorn sheep metapopulation. Journal of Wildlife Management 71:1080–1088.

Catry, I., T. Catry, M. Alno, A. A. A. Franco, and F. Moreia. 2016. Sexual and parent-offspring dietary segregation in a colonial raptor as revealed by stable isotopes. Journal of Zoology 299:58–67.

Catry, P., S. Bearhop, and M. Lecoq. 2007. Sex differences in settlement behaviour and condition of chiffchaffs Phylloscopus collybita at a wintering site in Portugal. Are females doing better? Journal of Ornithology 148:241–249.

Catry, P., R A. Phillips, and J. P. Croxall. 2005. Sexual segregation in birds: patterns, processes and implication for conservation. Pages 351–378 in K. E. Ruckstuhl and P. Neuhaus, editors. Sexual segregation in vertebrates: ecology of the two sexes. Cambridge University Press, Cambridge, United Kingdom.

Catry, T., J. A. Alves, J. A. Gill, T. Gunnarsson, and J. P. Grandeiro. 2012. Sex promotes spatial and dietary segregation in a migratory shorebird during the nonbreeding season. PLoS ONE 10:e33811.

Caughley, G. 1970. Eruption of ungulate populations, with emphasis on Himalayan thar in New Zealand. Ecology 51:53–72.

Caughley, G., and C. J. Krebs. 1983. Are big mammals simply little mammals writ large? Oecologia 59:7–17.

Ceballos, G., and P. R. Ehrlich. 2002. Mammal population losses and the extinction crisis. Science 296:904–907.

Ceballos, G., P. R. Ehrlich, and R. Dirzo. 2017. Biological annihilation via the ongoing sixth mass extinction signaled by vertebrate population losses and declines. Proceedings of the National Academy of Sciences of the United States of America 114:E6089–E6096.

Charles, W. N., D. McCowan, and K. East. 1977. Selection of upland swards by red deer (Cervus elaphus L.) on Rhum. Journal of Applied Ecology 145:55–64.

Chen, X., L. Vierling, E. Rowell, and T. Defelice. 2004. Using LIDAR and effective LAI data to evaluate IKONOS and LANDSAT 7 ETM+ vegetation cover estimates in a ponderosa pine forest. Remote Sensing of Environment 91:14–26.

Chesser, R. K. 1991a. Gene diversity and female philopatry. Genetics 127:437–447.

Chesser, R. K. 1991b. Influence of gene flow and breeding tactics on gene diversity within populations. Genetics 129:573–583.

Christianson, D., M. S. Becker, A. Brennan, S. Creel, E. Dröge, J. M'soka, T. Mukula, P. Schuette, D. Smit, and Fred Watson. 2018. Foraging investment in a long-lived herbivore and vulnerability to coursing and stalking predators. Ecology and Evolution 8:10147–10155.

Church, D. C. 1969. Digestive physiology and nutrition of ruminants. Volume 1. O.S.U. Bookstores, Corvallis, Oregon, USA.

Ciach, M., D. Wikar, M. Bylicka, and M. Bylicka. 2010. Flocking behavior and sexual segregation in black grouse Tetrao tetrix during the non-breeding period. Zoological Studies 49:453–460.

Cistrone, N. V., I. D. Salvo, A. Ariano, A. Fulco, and D. Russo. 2015. How to be a male at different elevations: ecology of intra-sexual segregation in the trawling bat Myotis daubentonii. PLoS ONE 10:e134573.

Ciuti, S., and M. Apollonio. 2008. Ecological sexual segregation in fallow deer (Dama dama): a multispatial and multitemporal approach. Behavioral Ecology and Sociobiology 62:1747–1759.

Ciuti, S., and M. Apollonio. 2016. Reproductive timing in a lekking mammal: male fallow deer getting ready for female estrus. Behavioral Ecology 27:1522–1532.

Ciuti, S., S. Davini, S. Luccarini, and M. Apollonio. 2004. Could the predation risk hypothesis explain large-scale sexual segregation in fallow deer (Dama dama)? Behavioral Ecology and Sociobiology 56:552–564.

Clark, B. L., S. L. Cox, K. M. Atkins, S. Bearhop, A. W. J. Bicknell, T. W. Bodey, I. R. Cleasby, W. J. Grecian, K. C. Hamer, B. R. Loveday, et al. 2021. Sexual segregation of gannet foraging over 11 years: isotopic differences remain stable. Marine Progress Ecology Series 661:1–16.

Clark, J. T., J. S. Horn, M. Hebblewhite, and A. D. Luis. 2021. Stochastic predation exposes prey to predator pit and local extinction. Oikos 130:300–309.

Clauss, M., A. Schwarm, S. Ortmann, W. J. Streich, and J. Hummel. 2007. A case of non-scaling in mammalian physiology? Body size, digestive capacity, food intake, and ingesta passage in mammalian herbivores. Comparative Biochemistry and Physiology A 148: 249–265.

Cleasby, I. R., E. D. Wakefield, T. W. Bodey, R. D. Davies, S. C. Patrick, J. Newton, S. C. Votier, S. Bearhop, and K. C. Hamer. 2015. Sexual segregation in a wide-ranging marine predator is a consequence of habitat selection. Marine Ecology Progress Series 518:1–12.

Clutton-Brock, T. H. 1977. Some aspects of intraspecific variation in feeding and ranging behaviour in primates. Pages 539–556 in T. H. Clutton-Brock, editor, Primate ecology. Cambridge University Press, Cambridge, United Kingdom.

Clutton-Brock, T. H. 1991. The evolution of parental care. Princeton University Press, Princeton, New Jersey, USA.

Clutton-Brock, T. H. 2016. Mammal societies. John Wiley and Sons, Oxford, United Kingdom.

Clutton-Brock, T. H., and S. D. Albon. 1989. Red deer in the highlands. BSP Professional Books, Oxford, United Kingdom.

Clutton-Brock, T. H., and P. H. Harvey. 1983. The functional significance of variation of body size among mammals. Special Publication of the American Society of Mammalogists 7:632–663.

Clutton-Brock, T. H., and G. A. Parker. 1992 Potential reproductive rates and the operation of sexual selection. Quarterly Review of Biology 67:437–456.

Clutton-Brock, T. H., S. D. Albon, and P. H. Harvey. 1980. Antlers, body size, and breeding group size in the Cervidae. Nature 285:565–567.

Clutton-Brock, T. H., F. E Guinness, and S. D. Albon. 1982a. Red deer: behavior and ecology of two sexes. University of Chicago Press, Chicago, Illinois, USA.

Clutton-Brock, T. H., G. R. Iason, S. D. Albon, and F. E. Guinness. 1982b. Effects of lactation on feeding behaviour and habitat use in wild red deer hinds. Journal of Zoology 198:227–236.

Clutton-Brock, T. H., G. R. Iason, and F. E. Guinness. 1987. Sexual segregation and density-related changes in habitat use in male and female red deer (Cervus elaphus). Journal of Zoology 211:275–289.

Coblentz, B. E. 1980. On the improbability of pursuit invitation signals in mammals. American Naturalist 115:438–442.

Coffin, A. W. 2007. From roadkill to road ecology: a review of the ecological effects of roads. Journal of Transport Geography 15:396–406.

Cohen, T. M., Y. Kiat, H. Sharon, and E. Levin. 2021. An alternative hypothesis for the evolution of sexual segregation in endotherms. Global Ecology and Biogeography 30:2420–2430.

Coltman, D. W., M. Festa-Bianchet, J. T. Jorgenson, and C. Strobeck. 2002. Age-dependent sexual selection in bighorn rams. Proceedings of the Royal Society of London Series B 269:165–172.

Conard, J. J., J. Serangeli, G. Bigga, and V. Rots. 2020. A 300,000-year-old throwing stick from Schöningen, northern Germany, documents the evolution of human hunting. Nature Ecology and Evolution 4:690–693.

Conde, D. A., F. Colchero, H. Zarza, N. L. Christensen Jr., J. O. Sexton, C. Manterola, C. Chávez, A. Rivera, D Azuara, and G. Ceballos. 2010. Sex matters: modeling male and female habitat differences for jaguar conservation. Biological Conservation 143:1980–1988.

Connell, J. H. 1980. Diversity and the coevolution of competitors, or the ghost of competition past. Oikos 35:131–138.

Conradt, L. 1998a. Could asynchrony in activity between the sexes cause intersexual social segregation in ruminants? Proceedings of the Royal Society of London B 265:1359 1363.

Conradt, L. 1998b. Measuring the degree of sexual segregation in group-living animals. Journal of Animal Ecology 67:217–226.

Conradt, L. 1999. Social segregation is not a consequence of habitat segregation in red deer and Soay sheep. Animal Behaviour 57:1151–1157.

Conradt, L. 2005. Definitions, hypotheses, models and measure in the study of animal segregation. Pages 11–32 in K. E. Ruckstuhl and P. Neuhaus, editors. Sexual segregation in vertebrates: ecology of the two sexes. Cambridge University Press, Cambridge, United Kingdom.

Conradt, L., and T. J. Roper. 2000. Activity synchrony and social cohesion: a fission-fusion model. Journal of Zoology 267:2213–2218.

Conradt, L., T. H. Clutton-Brock, and F. E. Guinness. 2000. Sex differences in weather sensitivity can cause habitat segregation: red deer as an example. Animal Behaviour 59:1049–1060.

Conradt, L., T. H. Clutton-Brock, and D. Thomson. 1999. Habitat segregation in ungulates: are males forced into suboptimal foraging habitats through indirect competition by females? Oecologia 119:376–377.

Conradt, L., I. J. Gordon, T. H. Clutton-Brock, and F. E. Guinness. 2001. Could the indirect competition hypothesis explain inter-sexual site segregation in red deer (*Cervus elaphus* L.)? Journal of Zoology 254:185–193.

Cook, J. G., B. K. Johnson, R. C. Cook, R. A. Riggs, T. Delcurto, L. D. Bryant, and L. L. Irwin. 2004. Effects of summer–autumn nutrition and parturition date on reproduction and survival of elk. Wildlife Monographs 155:1–61.

Cooper, J., P. Waser, E. Hellgren, T. Gabor, and J. A. DeWoody. 2011. Is sexual monomorphism a predictor of polyandry? Evidence for a social mammal, the collared peccary. Behavioral Ecology and Sociobiology 65:775–785.

Cooper, N. W., M. A. Thomas, and P. P. Marra. 2021. Vertical sexual habitat segregation in a wintering song bird. Ornithology 138:1–11.

Corlatti, L., B. Bassano, and S. Lovari. 2020. Weather stochasticity and alternative reproductive tactics in northern chamois, *Rupicapra rupicapra*. Biological Journal of the Linnean Society 130:359–364.

Corlatti, L., M. Caroli, V. Pietrocini, and S. Lovari. 2013. Rutting behaviour of territorial and non-territorial male chamois: is there a home advantage? Behavioural Processes 92:118–124.

Corti, P., and D. M. Shackleton. 2002. Relationship between predation-risk factors and sexual segregation in Dall's sheep (*Ovis dalli dalli*). Canadian Journal of Zoology 80:2108–2117.

Cosse, M., and S. González. 2013. Demographic characterization and social patterns of the Neotropical pampas deer. SpringerPlus 2:259.

Côté, S. D., J. A Schaefer, and F. Messier. 1997. Time budgets and synchrony of activities in muskoxen: the influence of sex, age, and season. Canadian Journal of Zoology 75:1628–1635.

Coulson, G., A. M. MacFarlane, S. E. Parsons, and J. Cutter. 2006. Evolution of sexual segregation in mammalian herbivores: kangaroos as marsupial models. Australian Journal of Zoology 54:217–224.

Couturier, S., S. D. Côté, J. Huot, and R. D. Otto. 2009. Body-condition dynamics in a northern ungulate gaining fat in winter. Canadian Journal of Zoology 87:367–378.

Craine, J. M. 2021. Seasonal patterns of bison diet across climate gradients in North America. Scientific Reports 11:6829.

Crampe, J.-P., J.-F. Gerar, M. Goulard, C. Milleret, G. Gonzalez, and R. Bon. 2021. Year-round sexual segregation in the Pyrenean chamois, a nearly monomorphic polygynous herbivore. Behavioural Processes 184:104300.

Cransac, N., J.-F. Gerard, M.-L. Maublanc, and D. Pépin. 1998. An example of segregation between age and sex classes only weakly related to habitat use in mouflon sheep (*Ovis gmelini*). Journal of Zoology 244:371–378.

Creel, S., and N. M. Creel. 2002. The African wild dog: behavior, ecology and conservation. Princeton University Press, Princeton, New Jersey, USA.

Croft, D. P., M. S. Botham, and J. Krause. 2004. Is sexual segregation in the guppy, *Poecilia reticulata*, consistent with the predation risk hypothesis? Environmental Biology of Fishes 71:127–133.

Croft, D. P., L. J. Morrell, A. S. Wade, C. Piyapong, C. C. Ioannou, J. R. Dyer, B. B. Chapman, W. Yan, and J. Krause. 2006. Predation risk as a driving force for sexual segregation: a cross-population comparison. American Naturalist 167:867–878.

Cryan, P. M., and B. O. Wolf. 2003. Sex differences in the thermoregulation and evaporative water loss of a heterothermic bat, *Lasiurus cinereus*, during its spring migration. Journal of Experimental Biology 206:3381–3390.

Cummings, J. L. 2019. Critical legislative and institutional underpinning of the North American model. Pages 56–72 in S. P. Mahoney and V. Geist, editors. The North American model of wildlife conservation. Johns Hopkins University Press, Baltimore, Maryland, USA.

Dale, B. W., L. G. Adams, and R. T. Bowyer. 1994. Functional response of wolves preying on barren-ground caribou in a multiple-prey ecosystem. Journal of Animal Ecology 63:644–652.

Dale, J., P. O. Dunn, J. Figuerola, T. Lislevand, T. Székely. and L. A. Whittingham. 2007. Sexual selection explains Rensch's rule of allometry for sexual size dimorphism. Proceedings of the Royal Society of London B 274: 2971–2979.

Dalmau, A., A. Ferret, J. L. R. de la Torre, and X. Manteca. 2013. Habitat selection and social behaviour in a Pyrenean chamois population (*Rupicapra pyrenaica pyrenaica*). Journal of Mountain Ecology 9:83–102.

Danilewicz, D. E. R. Secchi, P. H. Ott, I. B. Moreno, M. Bassoi, and M. Borges-Martins. 2009. Habitat use patterns of Franciscan dolphins (*Pontoporia blainvillei*) off southern Brazil in relation to water depth. Journal of the Marine Biological Association of the United Kingdom 89:943–949.

Darling, F. F. 1937. A herd of red deer. Oxford University Press, London, United Kingdom.

Darwin, C. R. 1859. On the origin of species by natural selection. Murray, London, United Kingdom.

Darwin, C. R. 1871. The descent of man, and selection in relation to sex. Murray, London, United Kingdom.

Darwin, C. R. 1874. The descent of man, and selection in relation to sex. Second edition. Murray, London, United Kingdom.

Davidson, A. D., K. T. Shoemaker, B. Weinstein, G. C. Costa, T. M. Brooks, G. Ceballos, V. C. Radeloff, C. Rondinini, and C. H. Graham. 2017. Geography of current and future global mammal extinction risk. PLoS ONE 12:e0186934.

de Albernaz, T., E. R. Secchi, L. R. de Oliveira, and S. Botta. 2016. Ontogenetic and gender-related variation in the isotopic niche within and between three species of fur seals (genus *Arctocephalus*). Hydrobiologia 787:123–139.

De'ath, G., and K. E. Fabricius. 2000. Classification and regression trees: a powerful yet simple technique for ecological data analysis. Ecology 81:3178–3192.

Debeffe, L., I. M. Rivrud, E. L. Meisingset, and A. Mysterud. 2019. Sex-specific differences in spring and autumn migration in a northern large herbivore. Scientific Reports 9:6137.

de Calesta, D. S. 1994. Effects of white-tailed deer on songbirds within managed forests in Pennsylvania. Journal of Wildlife Management 58:711–718.

DeCesare, N. J., M. Hebblewhite, H. S. Robinson, and M. Musiani. 2010. Endangered, apparently: the role of apparent competition in endangered species conservation. Animal Conservation 13:353–362.

Dechen Quinn, A. C., D. M. Williams, and W. F. Porter. 2013. Landscape structure influences space use by white-tailed deer. Journal of Mammalogy 94:398–407.

De Felipe, F., J. M. Reyes-González, R. Militão, V. C. Neves, J. Bried, D. Oro, R. Ramos, and J. González-Solís. 2019. Does sexual segregation occur during the nonbreeding period? A comparative analysis in spatial and feeding ecology of three *Calonectris* shearwaters. Ecology and Evolution 9:10145–10162.

Dehn, M. M. 1990. Vigilance for predators: detection and dilution effects. Behavioral Ecology and Sociobiology 26:337–342.

Delaney, D. M., and D. A. Warner. 2016. Age- and sex-specific variations in microhabitat use in a territorial lizard. Behavioral Ecology and Sociobiology 70:981–991.

de Lima, R. C., V. Franco-Trecu, D. G. Vales, P. Inchausti, E. R. Secchi, and S. Botta. 2019. Individual forging specialization and sexual niche segregation in South American fur seals. Marine Biology 166:32.

Dell'Apa, A., J. Cudney-Burch, D. G. Kimmel, and R. A. Rulifson. 2014. Sexual segregation of spiny dogfish in fishery-dependent surveys in Cape Cod, Massachusetts: potential management benefits. Transactions of the American Fisheries Society 143:833–844.

Demarais, S., and B. K. Strickland. 2011. Antlers. Pages 107–145 *in* D. G. Hewitt, editor. Biology and management of white-tailed deer. CRC Press, Boca Raton, Florida, USA.

Demment, M. W. 1982. The scaling of ruminoreticulum size with body-weight in East African ungulates. African Journal of Ecology 20:43–47.

Demment, M. W., and P. J. Van Soest. 1985. A nutritional explanation for body-size patterns of ruminant and nonruminant herbivores. American Naturalist 125:641–672.

DePerno, C. S., J. A. Jenks, and S. L. Griffin. 2003. Multidimensional cover characteristics: is variation in habitat selection related to white-tailed deer sexual segregation? Journal of Mammalogy 84:1316–1329.

De-Pin, L., M. B. Scott, X. Lin, and X Wen. 2013. Age-gender feeding differences in dwarf blue sheep, *Pseudois schaeferi* (Cetartiodactyla, Bovidae) magnified by the expansion of an invasive plant species. Mammalia 77:131–140.

Dinsmore, S. J., and D. H. Johnson. 2012. Population analysis in wildlife biology. Pages 349–380 *in* N. J. Silvy, editor. The wildlife techniques manual: research. Volume 1. Seventh edition. Johns Hopkins University Press, Baltimore, Maryland, USA.

Di Stefano, J., A. York, M. Swan, A. Greenfield, and G. Coulson. 2009. Habitat selection by the swamp wallaby (*Wallabia bicolor*) in relation to diel period, food and shelter. Austral Ecology 34:143–155.

Dixson, A. F. 2009. Sexual selection and the origins of human mating systems. Oxford University Press, Oxford, United Kingdom.

Domeier, M. L., and N. Nasby-Lucas. 2012. Sex-specific migration patterns and sexual segregation of adult white sharks, *Carcharodon carcharias*, in the Northeastern Pacific. Pages 133–146 *in* M. L. Domeier, editor. Global perspectives on the biology and life history of the white shark. CRC Press, Boca Raton, Florida, USA.

Donovan, V. M., S. P. H. Dwinnell, J. L. Beck, C. P. Roberts, J. G. Clapp, G. S. Hiatt, K. L. Monteith, and D. Twidwell. 2021. Fire-driven landscape heterogeneity shapes habitat selection of bighorn sheep. Journal of Mammalogy 102:757–771.

Dorst, J. 1969. A field guide to the larger mammals of Africa. Houghton Mifflin, Boston, Massachusetts, USA.

Drago, M., V. Franco-Trecu, L. Zenteno, D. Szteren, E. A. Crespo, F. R. Sapriza, L. de Oliveira, R. Machado, P. Inchausti, and L. Cardona. 2006. Sexual foraging segregation in South American sea lions increases during the pre-breeding period in the Rio de la Plata plume. Marine Ecology Progress Series 525:261–272.

Dressel, S., C. Sandström, and G. Ericsson. 2015. A meta-analysis of studies on attitudes toward bears and wolves across Europe 1976–2012. Conservation Biology 29:565–574.

Drymon, J. M., S. Dedman, J. T. Froeschke, E. A. Seubert, A. E. Kroetz, J. F. Mareska, and S. P. Powers. 2020. Defining sex-specific habitat suitability for a northern Gulf of Mexico shark assemblage. Frontiers in Marine Science 7:35.

Dunbar, R. I. M., and P. Dunbar. 1980. The pair bond in the klipspringer. Animal Behaviour 28:219–229.

Dunbar, R. I. M., and J. Shi. 2008. Sex differences in feeding activity results in sexual segregation of feral goats. Ethology114:444–451.

Duncan, P. 1991. Horses and grasses: the nutritional ecology of equids and their impact on the Camargue. Springer-Verlag, New York, New York, USA.

Duron, Q., J. E. Ménez, P. M. Vergara, G. E. Soto, M. Lizama, and R. Rozzi. 2017. Intersexual segregation in foraging microhabitat use by Magellanic woodpeckers (*Campephilus magellanicus*): seasonal and habitat effects at the world's southernmost forests. Austral Ecology 43:25–34.

Dussault, C., R. Courtois, J.-P. Ouellet, and I. Girard. 2005. Space use of moose in relation to food availability. Canadian Journal of Zoology 83:1431–1437.

du Toit, J. T. 1990. Feeding height stratification among African browsing ruminants. African Journal of Ecology 28:55–61.

du Toit, J. T. 1995. Sexual segregation in kudu: sex differences in competitive ability, predation risk or nutritional needs? South African Wildlife Research 25:127–132.

du Toit, J. T. 2005. Sex differences in the foraging ecology of large mammalian herbivores. Pages 36–52 *in* K. E. Ruckstuhl and P. Neuhaus, editors. Sexual segregation in vertebrates: ecology of the two sexes. Cambridge University Press, Cambridge, United Kingdom.

du Toit, J. T., and C. A. Yetman. 2005. Effects of body size on the diurnal activity budgets of African browsing ruminants. Oecologia 143:317–325.

Dyer, M. I., M. L. Turner, and T. R. Seastedt. 1993. Herbivory and its consequences. Ecological Applications 3:10–16.

Eberhardt, L. L. 2002. A paradigm for population analysis of long-lived vertebrates. Ecology 83: 2841–2854.

Eckrich, D. A., S. E. Albeke, E. A. Flaherty, R. T. Bowyer, and M. Ben-David. 2020. rKIN: kernel-based method for estimating isotopic niche size and overlap. Journal of Animal Ecology 89:757–771.

Edwards, J. 1983. Diet shifts in moose due to predator avoidance. Oecologia 60:185–189.

Eifler, D. A., M. A. Eifler, and E. N. Eifler. 2007. Habitat use and movement patterns in the graceful crag lizard, *Pseudocordylus capensis*. African Zoology 42:152–157.

Eisenberg, J. F. 1981. The mammalian radiations: an analysis of trends in evolution, adaptation, and behavior. University of Chicago Press, Chicago Illinois, USA.

Elliott Smith, E. A., S. D. Newsome, J. A. Estes, and M. T. Tinker. 2015. The cost of reproduction: differential resource specialization in female and male California sea otters. Oecologia 178:17–29.

Ellsworth, E., M. R. Boudreau, K. Nagy, J. L. Rachlow, and D. L. Murray. 2016. Differential sex related winter energetics in free-ranging snowshoe hares (*Lepus americanus*). Canadian Journal of Zoology 94:115–121.

Elson, L. T., F. E. Schwab, and N. P. P. Simon. 2007. Winter food habits of *Lagopus lagopus* (willow ptarmigan) as a mechanism to explain winter sexual segregation. Northeast Naturalist 14:89–98.

Emlen, S. T., and L. W. Oring. 1977. Ecology, sexual selection and the evolution of mating systems. Science 197:215–223.

Encarnãço, J. 2012. Spatiotemporal patterns of sexual segregation in a tree-dwelling temperate bat (*Myotis daubentonii*). Journal of Ethology 30:271–278.

Epps, C. W., P. J. Palsbøll, J. D. Wehausen, G. K. Roderick, R. R. Ramey II, and D. R. McCullough. 2005. Highways block gene flow and cause a rapid decline in genetic diversity of desert bighorn sheep. Ecology Letters 8:102–1038.

Errington, P. L. 1946. Predation and vertebrate populations. Quarterly Review of Biology 21: 144–177.

Estes, R. D. 1974. Social organization of the African Bovidae. Pages 166–205 *in* F. R. Walther and V. Geist, editors. The behaviour of ungulates and its relation to management. International Union for the Conservation of Nature and Natural Resources, Morges, Switzerland.

Estes, R. D. 1976. The significance of breeding synchrony in wildebeest. East African Wildlife Journal 14:135–152.

Estes, R. D., and R. K. Estes. 1979. The birth and survival of wildebeest calves. Zeitschrift für Tierpsychologie 50:45–95.

Estes, R. D., and J. Goddard. 1967. Prey selection and hunting behavior of the African wild dog. Journal of Wildlife Management 31:52–70.

Fairbairn, D. J. 1997. Allometry for sexual size dimorphism: pattern and process in the coevolution of body size in males and females. Annual Review of Ecology and Systematics 28:659–687.

Farmer, C. F., D. K. Person, and R. T. Bowyer. 2006. Risk factors and survivorship of black-tailed deer in a managed forest landscape. Journal of Wildlife Management 70:1403–1415.

Fancy, S. G., and R. G. White. 1985. Incremental cost of activity. Pages 145–159 *in* R. J. Hudson and R. G. White, editors. Bioenergetics of wild herbivores. CRC Press, Boca Raton, Florida, USA.

Fattorini, N., C. Brunetti, C. Baruzzi, G. Chiatante, S. Lovari, and F. Ferritti. 2019. Temporal variation in foraging activity and grouping patterns in a mountain-dwelling herbivore: environmental and endogenous drivers. Behavioural Processes 167:103909.

Fay, F. H. 1982. Ecology and biology of the Pacific walrus, *Odobenus rosmarus divergens* Illiger. North American Fauna 74:1–279.

Feldhamer, G. A., J. F Merritt, C. Krajewski, J. L. Rachlow, and K. M. Stewart. 2020. Mammalogy: adaptation, diversity, ecology. Fifth edition. Johns Hopkins University Press, Baltimore, Maryland, USA.

Ferrari, H., R. Rosá, P. Lanfranchi, and D. E. Ruckstuhl. 2010. Effect of sexual segregation on host-parasite interaction: model simulation for abomasal parasite dynamics in alpine ibex (*Capra ibex*). International Journal for Parasitology 40:1285–1293.

Ferretti, F., A. Costa, M. Corazza, V. Pietrocini, G. Cesaretti, and S. Lovari. 2014. Males are faster foragers than females: intersexual differences in foraging behaviour in the Apennine chamois. Behavioral Ecology and Sociobiology 68:1335–1344.

Festa-Bianchet, M. 1988. Seasonal range selection in bighorn sheep: conflicts between forage quality, forage quantity, and predator avoidance. Oecologia 75.580–586.

Festa-Bianchet, M., and S. D. Côté. 2008. Mountain goats: ecology, behavior, and conservation of an alpine ungulate. Island Press, Washington, DC, USA.

Fichter, E. 1974. On bedding behavior of pronghorn fawns. Pages 352–355 *in* V. Geist and F. Walther, editors. The behaviour of ungulates and its relation to management. New Series 24. International Union for the Conservation of Nature and Natural Resources, Morges, Switzerland.

Fieberg, J., and C. O. Kochanny. 2005. Quantifying home-range overlap: the importance of the utilization distribution. Journal of Wildlife Management 69:1346–1359.

Fitzgerald, A. E., and D. C. Waddington. 1979. Comparison of two methods of fecal analysis and herbivore diet. Journal of Wildlife Management 43:468–475.

Flook, D. R. 1970. A study of sex differential in the survival of wapiti. Canadian Wildlife Service Bulletin Report Series 11:1–71.

Flowerdew, J. R., and S. A. Ellwood. 2001. Impacts of woodland deer on small mammal ecology. Forestry 74:277–287.

Focardi, S., K. Farnsworth, B. M. Poli, M. P. Ponzetta, and A. Tinelli. 2003. Sexual segregation in ungulates:

individual behaviour and the missing link. Population Ecology 45:83–95.

Forchhammer, M. C., T. H. Clutton-Brock, J. Lindström, and S. D. Albon. 2001. Climate and population density induce long-term cohort variation in a northern ungulate. Journal of Animal Ecology 70:721–729.

Ford, A. T., J. R. Goheen, D. J. Augustine, M. F. Kinnaird, T. G. O'Brien, T. M. Palmer, R. M. Pringle, and R. Woodroff. 2015. Recovery of African wild dogs suppress prey but does not trigger a trophic cascade. Ecology 96:2575–2584.

Forero, M. G., J. González-Solis, K. A. Hobson, J. A. Donázar, M. Bertellotti, G. Blanco, and G. R. Bortolotti. 2005. Stable isotopes reveal trophic segregation by sex and age in the southern petrel in two different food webs. Marine Ecology Progress Series 296:107–113.

Forsyth, D. M., and P. Caley. 2006. Testing the irruptive paradigm of large herbivore dynamics. Ecology 87:297–303.

Fortin, D., M.-E., Fortin, H. L. Beyer, T. Duchesne, S. Courant, and K. Dancose. 2009. Group-size mediated habitat selection and group fusion-fission dynamics of bison under predation risk. Ecology 90:2480–2490.

Fowler, C. W. 1987. A review of density dependence in large mammal populations. Current Mammalogy 1:401–441.

Fox, M. W. 1984. The whistling hunters: field studies of the Asiatic wild dog (*Cuon alpinus*). State University of New York, Albany, New York, USA.

Francisci, F., S. Forardi, and L. Boitani. 1985. Male and female alpine ibex: phenology of space use, and herd size. Pages 124–133 *in* S. Lovari, editor. Management and biology of mountain ungulates. Croom Helm, London, United Kingdom.

Frank, D. A., R. S. Inouye, N. Huntly, G. W. Minshall, and J. E. Anderson. 1994. The biogeochemistry of a north-temperate grassland with native ungulates—nitrogen dynamics in Yellowstone National Park. Biogeochemistry 26:163–188.

Franklin, W. L. 1983. Contrasting sociologies of South American's wild camelids: the vicuña and guanaco. Pages 573–629 *in* J. F. Eisenberg and D. G. Kleiman, editors. Advances in the study of mammalian behavior. Special Publication of the American Society of Mammalogists 7. Stillwater, Oklahoma, USA.

Freeman, D. C., L. G. Klikoff, and K. T. Harper. 1976. Differential resource utilization by the sexes of dioecious plants. Science 193:597–599.

Freeman, E. D., R. T. Larsen, K. Clegg, and B. R. McMillian. 2013. Long-lasting effects of maternal condition in free-ranging cervids. PLoS ONE 8:e5873.

Freeman, E. D., R. T. Larsen, M. E. Peterson, C. R. Anderson Jr., K. R. Hersey, and B. R. McMillian. 2014.

Effects of male-biased harvest on mule deer: implications for rates of pregnancy, synchrony, and timing of parturition. Wildlife Society Bulletin 38: 806–811.

Frid, A. 1994. Observation on habitat use and social organization of a huemul *Hippocamelus bisulcus* coastal population in Chile. Biological Conservation 67:13–19.

Fritz, H., and A. Loison. 2006. Large herbivores across biomes. Pages 19–49 *in* K. Danell, P. Duncan, R. Bergström, and J. Pastor, editors. Large herbivore ecology, ecosystem dynamics and conservation. Cambridge University Press, Cambridge, United Kingdom.

Fryxell, J. M. 1991. Forage quality and aggregation by large herbivores. American Naturalist 138:478–498.

Fryxell, J. M., and A. R. E. Sinclair. 1988. Causes and consequences of migration by large herbivores. Trends in Ecology and Evolution 3:237–241.

Fryxell, J. M., D. J. T. Hussell, A. B. Lambert, and P. C. Smith. 1991. Time lags and population fluctuations in white-tailed deer. Journal of Wildlife Management 55:377–385.

Fulbright, T. E., and J. A. Ortega-S. 2006. White-tailed deer habitat: ecology and management on rangelands. Texas A&M University Press, College Station, Texas, USA.

Fulbright, T. E., W. F. Robbins, E. C. Hellgren, R. W. DeYoung, and I. D. Humphreys. 2001. Lack of diet partitioning by sex in reintroduced desert bighorn sheep. Journal of Arid Environments 48:49–52.

Fuller, T. K. 1989. Population dynamics of wolves in northcentral Minnesota. Wildlife Monographs 105:1–41.

Fury, C. A., K. E. Ruckstuhl, and P. L. Harrison. 2013. Spatial and social segregation patterns in Indo-Pacific bottlenose dolphins (*Tursiops aduncus*). PLoS ONE 8:e52987.

Gaillard, J.-M., M. Festa-Bianchet, N. G. Yoccoz, A. Loison, and C. Toïgo. 2000. Temporal variation in fitness components and population dynamics of large herbivores. Annual Review of Ecology and Systematics 31:367–393.

Galezo, A. A., E. Krzyszczyk, and J. Mann. 2018. Sexual segregation in Indo-Pacific bottlenose dolphins is driven by female avoidance of males. Behavioral Ecology 29:377–386.

Gallina, S., G. Sánchez-Rojas, A. Buenrostro-Silva, and C. A. Lópex-González. 2015. Comparison of faecal nitrogen concentration between sexes of white-tailed deer in a tropical dry forest in southern Mexico. Ethology Ecology and Evolution 27:103–115.

Gallina-Tessaro, S., L. A. Pérez-Solano, L. Garcia-Feria, G. Sánchez-Rojas, D. Hernández-Silva, and J. P. Esparza-Carlos. 2019. The mule deer of arid zones. Pages 347–369 *in* S. Gallina-Tessaro, editor. Ecology and conservation of tropical ungulates in Latin America. Springer International Publishing, New York, New York, USA.

Gantchoff, M., L. Conlee, and J. Belant. 2019. Conservation implication of sex-specific landscape suitability for a large generalist carnivore. Diversity and Distributions 25:1488–1496.

Garbarino, M., P. J. Weisberg, L. Bagnara, and C. Urbinanti. 2015. Sex-related spatial segregation along environmental gradients in the dioecious conifer, *Taxus baccata*. Forest Ecology and Management 358:122–129.

Garnick, S., J. Di Stefano, M. A. Elgar, and G. Coulson. 2014. Inter- and intraspecific effects of body size on habitat use among sexually-dimorphic macropodids. Oikos 123:984–992.

Garnick, S., J. Di Stefano, B. D. Moore, N. E. Davis, M. A. Elgar, and G. Coulson. 2018. Interspecific and intraspecific relationships between body mass and diet quality in a macropodid community. Journal of Mammalogy 99:428–439.

Gasaway, W. C., R. D. Boertje, D. V. Grandgaard, D. G. Kelleyhouse, R. O. Stephenson, and D. G. Larsen. 1992. The role of predation in limiting moose at low densities in Alaska and Yukon and implications for conservation. Wildlife Monographs 120:1–59.

Gastelum-Mendoza, F. I., I. Fernando, L. A. Tarango, L. Antonio, G. Olmos-Oropeza, J. Palaco-Núñez, D. Valdez-Zamudio, and R. Noriega-Valdez. 2021. Diet and sexual segregation of bighorn sheep (*Ovis canadensis mexicana* Merriam) in Sonora, Mexico. Argo Productividad 14:2043.

Gates, C., and R. J. Hudson. 1978. Energy costs of locomotion in wapiti. Acta Theriologica 23:365–370.

Gaudin, S., E. Chaillou, F. Cornilleau, C. Mossu, X. Bovin, and R. Nowak. 2015. Daughters are more strongly attached to their mother than sons: a possible mechanism for early social segregation. Animal Behaviour 102:33–43.

Gause, G. J. 1934. The struggle for existence. Williams and Wilkins, Baltimore, Maryland, USA.

Gaynor, K. M., J. S. Brown, A. D. Middleton, M. E. Power, and J. S. Brashares. 2019. Landscapes of fear: spatial patterns of risk perception and response. Trends in Ecology and Evolution 34:355–358.

Gedir, J. V., J. W. Cain III, G. Harris, and T. T. Turnball. 2015. Effects of climate change on long-term population growth of pronghorn in an arid environment. Ecosphere 6:art189.

Gedir, J. V., J. W. Cain III, T. L. Swetham, P. R. Krausman, and J. R. Morgart. 2020. Extreme drought and adaptive resources selection by a desert mammal. Ecosphere 11:e03175.

Geist, V. 1966. The evolution of horn-like organs. Behaviour 27:175–214.

Geist, V. 1971. Mountain sheep: a study in behavior and evolution. The University of Chicago Press, Chicago, Illinois, USA.

Geist, V. 1982. Adaptive behavioral strategies. Pages 291–277 in J. W. Thomas and D. E. Toweill, editors. Elk of North America: ecology and management. Stackpole Books, Harrisburg, Pennsylvania, USA.

Geist, V. 1988. How markets for wildlife meat and parts, and the sale of hunting privileges, jeopardize wildlife conservation. Conservation Biology 2:15–26.

Geist, V., and M. Bayer. 1988. Sexual dimorphism in Cervidae and its relationship to habitat. Journal of Zoology 214:45–53.

Geist, V., and P. T. Bromley. 1978. Why deer shed antlers. Zeitschrift für Säugetierkunde 45:223–231.

Geist, V., and R. G. Petocz. 1977. Bighorn sheep in winter: do rams maximize reproductive fitness by spatial and habitat segregation from ewes? Canadian Journal of Zoology 55:1802–1810.

Gerard, J.-F., and C. Richard-Hansen. 1992. Social affinities as the basis of the social organization of Pyrenean chamois (Rupicapra pyrenaica) population in an open mountain range. Behavioural Processes 28:111–122.

Germonpré, M. 2004. The influence of climate on sexual segregation and cub mortality in Peniglacial cave bear. Pages 51–63 in R. C. G. M. Lauwerier and I. Plug, editors. The future from the past: archaeozoology in wildlife conservation and heritage management. Oxbow Books, Oxford, United Kingdom.

Gese, E., P. Terletzky, S. M. C. Cavalcanti, and C. M. U. Neal. 2018. Influence of behavioral state, sex, and season on resource selection by jaguars (Panthera onca): always on the move. Ecosphere 9:art.E02341.

Gilbert, S. L., K. J. Hundertmark, M. S. Lindberg, D. K. Person, and M. S. Boyce. 2020. The importance of environmental variability in transient population dynamics for a northern ungulate. Frontiers in Ecology and Evolution 8:531027.

Ginnett, T. F., and M. W. Demment. 1997. Sex differences in giraffe foraging behavior at two spatial scales. Oecologia 110:291–300.

Goble, D. D., J. A. Wiens, J. M. Scott, T. D. Male, and J. A. Hall. 2012. Conservation-reliant species. BioScience 62:869–873.

Gomez, J. J., and M. H. Cassini. 2014. Analysis of environmental correlates of sexual segregation in northern elephant seals using species distribution models. Marine Biology 161:481–487.

González-Solís, J., and J. P. Croxall. 2005. Differences in foraging behaviour and feeding ecology in giant petrels. Pages 92–111 in K. E. Ruckstuhl and P. Neuhaus, editors. Sexual segregation in vertebrates: ecology of the two sexes. Cambridge University Press, Cambridge, United Kingdom.

Gordon, I. J., and A. W. Illius. 1988. Incisor arcade structure and diet selection in ruminants. Functional Ecology 2:15–22.

Gordon, I. J., and A. W. Illius. 1996. The nutritional ecology of African ruminants: a reinterpretation. Journal of Animal Ecology 65:18–28.

Gordon, I. J., and H. H. T. Prins, editors. 2019. The ecology of browsing and grazing II. Springer Nature, Cham, Switzerland.

Gordon, M. 1977. Animal physiology: principles and adaptations. McMillian, New York, New York, USA.

Gosling, L. M. 1986. The evolution of the mating strategies in male antelopes. Pages 244–281 in D. I. Rubenstein and R. W. Wrangham, editors. Ecological aspects of social evolution: birds and mammals. Princeton University Press, Princeton, New Jersey, USA.

Granroth-Wilding, H. M. V., and R. A. Phillips. 2019. Segregation in space and time explains the coexistence of two sympatric sub-Antarctic petrels. Ibis 161:101–116.

Gray, D. R. 1987. The muskoxen of Polar Bear Pass. Fitzhenry and Whiteside, Markham. Ontario, Canada.

Green, W. C., and A. Rothstein. 1993. Asynchronous parturition in bison: implications for the hider-follower dichotomy. Journal of Mammalogy 74:920–925.

Gregory, A. J., M. A. Lung, T. M. Gehring, and B. J. Swanson. 2009. The importance of sex and spatial scale when evaluating sexual segregation by elk in Yellowstone. Journal of Mammalogy 90:971–979.

Griffiths, S. W., J. E. Orpwood, A. F. Ojanguren, J. D. Armstrong, and A. E. Magurran. 2014. Sexual segregation in monomorphic minnows. Animal Behaviour 88:7–12.

Grignolio, S., F. Brivio, N. Sica, and M. Apollonio. 2019. Sexual differences in behavioural response in variation in predation risk. Ethology 125:603–612.

Grignolio, S., I. Rossi, B. Bassano, M. Apollonio. 2007. Predation risk as a factor affecting sexual segregation in alpine ibex. Journal of Mammalogy 88:1488–1497.

Groff, L. A., J. K. Calhoun, and C. S. Loftin. 1981. Terrestrial activity and summer home range of the mole salamander (Ambystoma talpoideum). Canadian Journal of Zoology 59:315–322.

Gross, J. E. 1998. Sexual segregation in ungulates: a comment. Journal of Mammalogy 79:1404–1409.

Gross, J. E., P. U. Alkon, and M. W. Demment. 1995. Grouping patterns and spatial segregation by Nubian ibex. Journal of Arid Environments 30:423–439.

Grovenburg, T. W., J. A. Jenks, C. N. Jacques, R. W. Klaver, and C. C. Swanson. 2009. Aggressive defensive behavior by free-ranging white-tailed deer. Journal of Mammalogy 90:1218–1223.

Groves, C., and P. Grubb. 2011. Ungulate taxonomy. Johns Hopkins University Press, Baltimore, Maryland, USA.

Guernsey, N. C., K. A. Lohse, and R. T. Bowyer. 2015. Rates of decomposition and nutrient release of herbivore inputs are driven by habitat microsite characteristics. Ecological Research 30:951–961.

Guilhem, C., E. Bideau, J.-F. Gerard, M.-L. Maublanc, and D. Pépin. 2006. Early differentiation of male and female interactive behaviour as a possible mechanism for sexual segregation in mouflon sheep (*Ovis gmelini*). Applied Animal Behaviour Science 98:54–69.

Gunn, A. 1992. Differences in the sex and age composition of two muskox populations and implications for male breeding strategies. Rangifer, 12:17–19.

Guthrie, R. D. 1971. A new theory of mammalian rump patch evolution. Behaviour 38:132–145.

Guthrie, R. D. 1990. Frozen fauna of the mammoth steppe: the story of Blue Babe. University of Chicago Press, Chicago, Illinois, USA.

Guthrie, R. D. 2000. Paleolithic art as a resource in artiodactyl paleobiology. Pages 96–127 in E. S. Vrba and G. B. Schaller, editors. Antelopes, deer, and relatives: fossil record, behavioral ecology, systematics, and conservation. Yale University Press, New Haven, Connecticut, USA.

Hairston, N. G., F. E. Smith, and L. B. Slobodkin. 1960. Community structure, population control, and competition. American Naturalist 94:421–425.

Hall, L. S., P. R. Krausman, and M. L. Morrison. 1997. The habitat concept and a plea for standard terminology. Wildlife Society Bulletin 25:173–182.

Halley, D. J., and M. Mari. 2004. Dry season social affiliation of African buffalo bulls at the Chobe riverfront, Botswana. South African Journal of Wildlife Research 34:105–111.

Hamilton, W. D. 1971. Geometry for selfish herd. Journal of Theoretical Biology, 31:295–311.

Han, L., D. Blank, M. Wang, A. Alves da Silva, W. Yang, K. Ruckstuhl, and J. Alves. 2020a. Diet differences between males and females in sexually dimorphic ungulates: a case study on Siberian ibex. European Journal of Wildlife Research 66:55.

Han, L., D. Blank, M. Wang, and W. Yang. 2020b. Vigilance behaviour in Siberian ibex (*Capra sibirica*): effects of group size, group type, sex and age. Behavioral Processes 170:104021.

Han, L., D. Blank, M. Wang, W. Yang, A. Alves de Silva, and J. Alves. 2019. Grouping patterns and social organization in Siberian ibex (*Capra sibirica*): feeding strategy matters. Folia Zoologica 68:35–42.

Han, L., Z. Wang, D. Blank, M. Wang, and W. Yang. 2021. Different environmental requirements of female and male Siberian ibex, *Capra sibirica*. Scientific Reports 11:6064.

Hanley, T. A. 1980. The nutritional basis for food selection by ungulates. Journal of Range Management 35:146–151.

Harder, L. D., S. C. H. Barrett, and W. W. Cole. 2000. The mating consequences of sexual segregation within inflorescences of flowering plants. Proceedings of the Royale Society of London B 267:315–320.

Hardin, G. 1960. The competitive exclusion principle. Science 131:1292–1297.

Harris, S., A. R. Rey, R. H. Phillips, and F. Quintana. 2013. Sexual segregation in timing of foraging by imperial shag (*Phalacrocorax atriceps*): is it always ladies first? Marine Biology 160:1249–1258.

Harrison, M. J. S. 1983. Age and sex differences in the diet and feeding strategies of the green monkey, *Cercopithecus sabaeus*. Animal Behaviour 31:96–977.

Harrison, R. D., R. Sreekar, J. F. Brodie, S. Brook, M. Luskin, H. O'Kelly, M. Rao, B. Scheffers, and N. Velho. 2016. Impacts of hunting on tropical forests in Southeast Asia. Conservation Biology 30:972–981.

Hart, E. E., J. Fennessy, E. Wells, and S. Ciuti. 2021. Seasonal shifts in sociosexual behaviour and reproductive phenology in giraffe. Behavioral Ecology and Sociobiology 75:15.

Hartwell, K. S., H. Notman, C. Bonefant, and M. S. Pavelka. 2014. Assessing the occurrence of sexual segregation in spider monkeys (*Ateles geoffroyi yucatanensis*), its mechanisms and function. International Journal of Primatology 35:425–444.

Hartwell, K. S., H. Notman, U. Kaibitzer, C. A. Chapman, M. M. S. M. Pavelka. 2021. Fruit availability has a complex relationship with fission-fusion dynamics in spider monkeys. Primates 62:165–175.

Hartwell, K. S., H. Notman, and M. S. M. Pavelka. 2018. Seasonal and sex differences in the fission-fusion dynamics of spider monkeys (*Ateles geoffroyi yucatanesis*) in Belize. Primates 59:531–539.

Harvey, B. D., K. N. Vanni, D. M. Shier, and G. F. Grether. 2017. Experimental test of mechanisms underlying sexual segregation at communal roosts of harvestmen (*Prionostemma* spp.). Ethology 123:516–525.

Haugen, J. B., T. H. Curtis, P. G. Ferandes, K. A. Sosebee, and P. J. Rago. 2017. Sexual segregation of spiny dogfish (*Squalus acanthias*) off the northeastern United States: implications for a male-directed fishery. Fisheries Research 193:121–128.

Haus, J. M., S. L. Webb, B. K. Strickland, K. P. McCarthy, J. E. Rogerson, and J. L. Bowman. 2020. Individual heterogeneity in resource selection has implications for mortality risk in white-tailed deer. Ecosphere 11:e03064.

Hawkins, E. R., L. Pogson-Manning, C. Jaehnichen, and J. J. Meager. 2020. Social dynamics and sexual segregation of Australian humpback dolphins (*Sousa sahulensis*) in Moreton Bay, Queensland. Marine Mammal Science 36:500–521.

Hay, C. T., P. C. Cross, and P. J. Funston. 2008. Trade-offs of predation and foraging explain sexual segregation in African buffalo. Journal of Animal Ecology 77:850–858.

Heady, H. F. 1975. Rangeland management. McGraw-Hill, New York, New York, USA.

Heard, D. C. 1992. The effect of wolf predation and snow cover on musk-ox group size. American Naturalist 139:190–204.

Hedd, A., W. A. Motevecchi, R. A. Phillips, and D. A. Fifield. 2014. Seasonal sexual segregation by monomorphic sooty shearwaters *Puffinus griseus* reflects different reproductive roles during the pre-laying period. PLoS ONE 9:e85572.

Hedrick, A. V., and E. J. Temeles. 1989. The evolution of sexual dimorphism in animals: hypotheses and tests. Trends in Ecology and Evolution 4:136–138.

Heffelfinger, J. R., and S. P. Mahoney. 2019. Hunting and vested interests as the spine of the North American model. Pages 83–94 in S. P. Mahoney and V. Geist, editors. The North American model of wildlife conservation. Johns Hopkins University Press, Baltimore, Maryland, USA.

Heffelfinger, L. J., K. M. Stewart, A. P. Bush, J. S. Sedinger, N. W. Darby, and V. C. Bleich. 2017. Timing of precipitation in an arid environment: effects on population performance of a large herbivore. Ecology and Evolution 2017:1–12.

Heffelfinger, L. J., K. M. Stewart, K. T. Shoemaker, N. W. Darby, and V. C. Bleich. 2020. Balancing current and future reproductive investment: variation in resource selection during stages of reproduction in a long-lived herbivore. Frontiers in Ecology and Evolution 8:163.

Hefty, K. L., and K. M. Stewart. 2018. Novel location data reveal spatiotemporal strategies used by a central-place forager. Journal of Mammalogy 99:333–340.

Helle, T. 1979. Sex segregation during calving and summer period in wild forest reindeer (*Rangifer tarandus fennicus* Lönn.) in eastern Finland with special reference to habitat requirements and dietary preferences. Proceedings of the International Reindeer and Caribou Symposium 2:508–518.

Heurich, M., K. Seis, H. Küchenhoff, J. Müller, E. Belotti, L. Vufka, and B. Woelfing. 2016. Selective predation of a stalking predator on ungulate prey. PLoS ONE 11:e0158449.

Hill, D. A., and M. W. Ridley. 1987. Sexual segregation in winter, spring dispersal and habitat use in the pheasant (*Phasianus colchicus*). Journal of Zoology 212:657–668.

Hillen, J., T. Kaster, J. Pahle, A. Kieffer, O. Elle, E. M. Grieber, and M. Veith, 2011. Sex-specific habitat selection in an edge habitat specialist, the western barbastelle bat. Annales Zoologici Fennici 48:180–190.

Hirth, D. H. 1977. Social behavior of white-tailed deer in relation to habitat. Wildlife Monographs 53:1–55.

Hirth, D. H., and D. R. McCullough. 1977. Evolution of alarm signals in ungulates with special reference to white-tailed deer. American Naturalist 111:31–42.

Hobbs, N. T., and D. M. Swift. 1985. Estimates of habitat carrying capacity incorporating explicit nutritional constraints. Journal of Wildlife Management 49:814–822.

Hochkirch, A., J. Groning, and S. Krause. 2007. Intersexual niche segregation in Cepero's ground-hopper, *Tetrix ceperoi*. Evolutionary Ecology 21:727–738.

Hodgman, T. P., and R. T. Bowyer. 1986. Fecal crude protein relative to browsing intensity by white-tailed deer on wintering areas in Maine. Acta Theriologica 31:347–353.

Hofmann, R. R. 1985. Digestive physiology of the deer: their morphophysiological specialization and adaptation. Pages 393–407 in P. F. Fennessy and K. R. Drew, editors. Biology of deer production. The Royal Society of New Zealand Bulletin 22. Wellington, New Zealand.

Hofmann, R. R. 1988. Anatomy of the gastro-intestinal tract. Pages 14–43 in D. C. Church, editor. The ruminant animal: digestive physiology and nutrition. Prentice Hall, Englewood Cliffs, New Jersey, USA.

Hoglander, C., B. G. Dickson, S. S. Rosenstock, and J. J. Anderson. 2015. Landscape models of space use by desert bighorn sheep in the Sonoran Desert of southwestern Arizona. Journal of Wildlife Management 79:77–91.

Holling, C. S. 1959. The component of predation as revealed by a study of small-mammal predation of the European pine sawfly. Canadian Entomologist 91:293–320.

Holmes, E. 1988. Foraging behaviors among different age and sex classes of Rocky Mountain goats. Biennial Symposium on Northern Wild Sheep and Goat Council Proceedings 6:13–25.

Holt, R. E., A. Foggo, F. C. Neat, and K. L. Howell. 2013. Distribution patterns and sexual segregation in chimaeras: implications for conservation and management. ICES Journal of Marine Science 70:1198–1205.

Hudson, R. J., and R. G. White, editors. 1985. Bioenergetics of wild herbivores. CRC Press, Boca Raton, Florida, USA.

Hull, D. B. 1964. Hounds and hunting in ancient Greece. University of Chicago Press, Chicago, Illinois, USA.

Husek, J., M. Panek, and P. Tryjanowski. 2015. Predation risk drives habitat-specific sex ratio in a monomorphic species, the brown hare (*Lepus europaeus*). Ethology 121:593–600.

Hutchinson, G. E. 1978. An introduction to population ecology. Yale University Press, New Haven, Connecticut, USA.

Ibáñez, C., A. Guillén, P. T. Agiree-Mendi, J. Juste, G. Schreur, A. E. Cordero, and A. G. Popa-Lisseanu. 2009. Sexual segregation in Iberian noctule bats. Journal of Mammalogy 90:235–243.

Ihl, C., and R. T. Bowyer. 2011. Leadership in mixed-sex groups of muskoxen during the snow-free season. Journal of Mammalogy 92:819–827.

Illius, A. W., and I. J. Gordon. 1987. The allometry of food intake in grazing ruminants. Journal of Animal Ecology 56:989–999.

Illius, A. W., and I. J. Gordon. 1991. Modeling the nutritional ecology of ungulates herbivores: evolution of body size and competitive interactions. Oecologia 89:428–434.

Isaac, N. J., S. T. Turvey, B. Collen, C. Waterman, and J. E. Baillie. 2007. Mammals on the edge: conservation priorities based on threat and phylogeny. PLoS ONE 2:e296.

Istvanko, D. R., T. S. Risch, and V. Rolland. 2016. Sex-specific forging habits and roost characteristics of *Nycticeius humeralis* in north-central Arkansas. Journal of Mammalogy 97:1336–1335.

Isvaran, K. 2005. Variation in male mating behaviour within ungulate populations: patterns and processes. Current Science 89:1192–1199.

Isvaran, K. 2020. Lek territory size and the evolution of leks: a model and a test using an ungulate with a flexible mating system. Frontiers in Ecology and Evolution 8:539061.

Jacobsen, N. K. 1979. Alarm bradycardia in white-tailed deer (*Odocoileus virginianus*). Journal of Mammalogy 60:343–349.

Jacques, C. N., J. A. Jenks, T. W. Grovenburg, and R. W. Klaver. 2015. Influence of habitat and intrinsic characteristics on survival of neonatal pronghorn. PLoS ONE 10:e144026.

Jaeggi, A. V., M. I. Miles, M. Festa-Bianchet, C. Schradin, and L. D. Hayes. 2020. Variable social organization is ubiquitous in Artiodactyla and probably evolved from pair-living ancestors. Proceedings of the Royal Society of London B 287:20200035.

Jakimchuk, R. D., S. H. Ferguson, and L. G. Sopuck, 1987. Differential habitat use and sexual segregation in the Central Arctic caribou herd. Canadian Journal of Zoology 65:534–541.

Janis, C. M. 1982. Evolution of horns in ungulates: ecology and paleoecology. Biological Reviews of the Cambridge Philosophical Society 57:261–318.

Jarman, M. V. 1979. Impala social behaviour, territory, hierarchy, mating, and the use of space. Supplements to Journal of Comparative Ethology 21:1–92.

Jarman, P. 1983. Mating system and sexual dimorphism in large terrestrial, mammalian herbivores. Biological Reviews 58:485–520.

Jarman, P. J. 1974. The social organization of antelope in relation to their ecology. Behaviour 48:215–266.

Jenks, J. A. 2018. Mountain lions of the Black Hills: history and ecology. Johns Hopkins University Press, Baltimore, Maryland, USA.

Jenks, J. A., D. M. Leslie Jr., R. L. Lochmiller, and M. A. Melchiors. 1994. Variation in gastrointestinal characteristic of male and female white-tailed deer: implication for resource partitioning. Journal of Mammalogy 75:1045–1053.

Jenks, J. A., W. P. Smith, and C. S. DePerno. 2002. Maximum sustained yield harvest versus trophy management. Journal of Wildlife Management 66:528–535.

Jesmer, B. R., M. J. Kauffman, A. B. Courtemanch, S. Kilpatrick, T. Thomas, J. Yost, K. L. Monteith, and J. R. Goheen. 2021. Life-history theory provides a framework for detecting resource limitation: a test of the nutritional buffer hypothesis. Ecological Applications 31:e02299.

Jhala, Y. V., and D. Isvaran. 2016. Behavioural ecology of a grassland antelope, the blackbuck *Antilope cervicapra*: linking habitat, ecology, and behaviour. Pages 151–176 *in* F. W. Ahrestani and M. Sankaran, editors. The ecology of large herbivores in South and Southeast Asia. Springer, New York, New York, USA.

Jiang, Z. 2007. Sexual segregation in Tibetan gazelle: a test of the activity budget hypothesis. Journal of Zoology 74:327–331.

Jiang, Z., S. Hamasaki, S. Takatsuki, M. Kishimoto, and M. Katahara. 2008. Seasonal sexual variation in the site and gastrointestinal features of the sika deer in western Japan: implications for the feeding strategy. Zoological Science 26:691–696.

Jiang, Z., B. Liu, Y. Zeng, G. Hang, and H. Hu. 2000. Attracted by the same sex, or repelled by the opposite sex? Sexual segregation in Père David's deer. Chinese Scientific Bulletin 45:485–491.

Jiménez, S., A. Domingo, A. Brazerio, and O. Defeo. 2017. Sexual size dimorphism, spatial segregation and sex-biased bycatch of southern and northern royal albatrosses in pelagic longline fisheries. Antarctic Science 29:147–154.

Johansson, W., G. Koehler, G. R. Rauset, G. Samelius, H. Andrén, C. Mishra, P. Lhagvasuren, R. McCarthy, and M. Low. 2018. Sex-specific seasonal variation in puma and snow leopard home range utilization. Ecosphere 9:e02371.

Johnson, D. H. 1980. The comparison of usage and availability measurements for evaluating resource preference. Ecology 61:65–71.

Johnson, H. E., D. D. Gustine, T. S. Golden, L. G. Adams, L. S. Aprrett, E. A. Lenart, and P. S. Barboza. 2018. NDVI exhibits mixed success in predicting spatiotemporal variation in caribou summer forage quality and quality. Ecosphere 9:e02461.

Johnson, H. E., M. Hebblewhite, T. R. Stephenson, D. W. German, B. M. Pierce, and V. C. Bleich. 2013. Evaluating apparent competition in limiting the recovery of an endangered ungulate. Oecologia 171:295–307.

Joly, K., E. Gurare, M. S. Sorum, P. Kaczensky, M. D. Cameron, A. F. Jakes, B. L. Borg, D. Nandintsetseg, J. Grant, C. Hopcraft, et al. 2019. Longest terrestrial migrations and movements around the world. Scientific Reports 9:art.15333.

Joly, K., M. S. Sorum, T. Craig, and E. L. Julianus. 2016. The effects of sex, terrain, wildlife, winter severity, and maternal status on habitat selection by moose in north central Alaska. Alces 52:101–115.

Jones, A. G., J. R. Arguello, and S. J. Arnold. 2002. Validation of Bateman's principles: a genetic study of sexual selection and mating patterns in the rough-skinned newt. Proceedings of the Royal Society of London B 269:2533–2539.

Jones, C. B. 2014. The evolution of mammalian sociality in an ecological perspective. Springer Brief, Springer, Cham, Switzerland.

Jones, H., P. J. Pekins, L. E. Kantar, M. O'Neil, and D. Ellingwood. 2017. Fecundity and summer calf survival of moose during 3 successive years of winter tick epizootics. Alces 53:85–98.

Jones, K. A., N. Ratcliffe, S. C. Votier, J. Newton, J. Forcada, J. Dickens, G. Stowasser, and I. J. Staniland. 2020a. Intra-specific niche partitioning in Antarctic fur seals, Arctocephalus gazelle. Scientific Reports 10:art.3238.

Jones, K. A., H. Wood, J. P. Ashburner, F. Forcada, N. Ratcliffe, S. C. Votier, and I. J. Staniland. 2020b. Risk exposure trade-offs in the ontogeny of sexual segregation in Antarctic fur seal pups. Behavioral Ecology 31:719–730.

Jönsson, K. 1997. Capital and income breeding as alternative tactics of resource use in reproduction. Oikos 78:57–66.

Journeaux, K. L., P. C. Gardner, H. Y. Lim, J. G. E. Wern, and B. Goossens. 2018. Herd demography, sexual segregation and the effects of forest management on Bornean banteng Bos javanicus lowi in Sabah, Malaysian Borneo. Endangered Species Research 35:141–157.

Kagima, B., and W. S. Fairbanks. 2013. Habitat selection and diet composition of reintroduced native ungulates in a fire-managed tallgrass prairie reconstruction. Ecological Restoration 31:79–88.

Kalcounis, M. C., and R. M. Brigham. 1995. Intraspecific variation in wing loading affects habitat use by little brown bats. (Myotis lucifugus). Canadian Journal of Zoology 73:89–95.

Kamler, J. F., B. Jędrzejewska, and W. Jędrzejewska. 2007. Activity patterns of red deer in Białowieża National Park, Poland. Journal of Mammalogy 88:508–514.

Kasozi, H., and R. A. Montgomery. 2020. Variability in the estimation of ungulate group sizes complicates ecological inference. Ecology and Evolution 10:6881–6889.

Katsikaros, K., and R. Shine. 1997. Sexual dimorphism in the tusked frog, Adelotus brevis (Anura: Myobatrachidae): the roles of natural and sexual selection. Biological Journal of the Linnean Society 60:39–51.

Katsis, L. K. D., D. M. Linton, and D. W. Macdonald. 2021. The effect of group size, reproductive condition and the time period on sexual segregation patterns in three vespertilionid bat species. Journal of Zoology 313:135–144.

Kauffman, M. J., J. F. Brodie, and E. S. Jules. 2010. Are wolves saving Yellowstone's aspen? A landscape-level test of a behaviorally mediated trophic cascade. Ecology 91:2742–2755.

Kawamura, K., T. Akiyama, H. Yokota, M. Tsutsumi, T. Yasuda, G. Watanabe, G. Wang, and S. Wang. 2005. Monitoring of forage conditions with MODIS imagery in the Xilingol steppe, Inner Mongolia. International Journal of Remote Sensing 26:1421–1436.

Keech, M. A., R. D. Boertje, R. T. Bowyer, and B. W. Dale. 1999. Effects of birth weight on growth of young moose: do low-weight neonates compensate? Alces 35:51–57.

Keech, M.A., R. T. Bowyer, J. M. Ver Hoef, R. D. Boertje, B. W. Dale, and T. R. Stephenson. 2000. Life-history consequences of maternal condition in Alaskan moose. Journal of Wildlife Management 64:450–462.

Keehner, J. R., R. B. Wielgus, B. T. Malezke, and M. E. Swanson. 2015. Effects of male targeted harvest regime on sexual segregation in mountain lions. Biological Conservation 192:42–47.

Keesing, F., and T. Crawford. 2001. Impacts of density and large mammals on space use by the pouched mouse (*Saccostomus mearnsi*) in central Kenya. Journal of Tropical Ecology 17:465–472.

Kernaléguen, L., J. P. Y. Arnould, C. Guinet, B. Cazelles, P. Richard, and Y. Cherel. 2016. Early-life sexual segregation: ontogeny of isotopic niche differentiation in the Antarctic fur seal. Scientific Reports 6:33211.

Kernaléguen, L., B. Cazelles, J. P. Amould, P. Richard, C. Guinet, and Y. Cherel. 2012. Long-term species, sexual and individual variations in foraging strategies of fur seals revealed by stable isotopes in whiskers. PLoS ONE 7:e32916.

Kernaléguen, L., Y., Cherel, T. C. Knox, and A. M. M. Baylis. 2015. Sexual niche segregation and gender-specific individual specialization in a highly dimorphic marine mammal. PLoS ONE 10:e0133018.

Kerr, N. L. 1998. HARKing: hypothesizing after results are known. Personality and Social Psychology Review 2:196–217.

Ketchum, J. T., F. Galván-Magaña, and A. P. Klimley. 2013. Segregation and foraging ecology of whale sharks, *Rhincodon typus*, in the southwest Gulf of California. Environmental Biology of Fishes 96:779–795.

Kie, J. G. 1999. Optimal foraging and risk of predation: effects on behavior and social structure in ungulates. Journal of Mammalogy 80:1114–1129.

Kie, J. G., and R. T. Bowyer. 1999. Sexual segregation in white-tailed deer: density-dependent changes in use of space, habitat selection, and dietary niche. Journal of Mammalogy 80:1004–1020.

Kie, J. G., and M. White. 1985. Population dynamics of white-tailed deer (*Odocoileus virginianus*) on the Welder Wildlife Refuge, Texas. Southwestern Naturalist 30:105–118.

Kie, J. G., R. T. Bowyer, M. C. Nicholson, B. B. Boroski, and E. R. Loft. 2002. Landscape heterogeneity at differing scales: effects on spatial distribution of mule deer. Ecology 83:530–544.

Kie, J. G., R. T. Bowyer, and K. M. Stewart. 2003. Ungulates in western forests: habitat requirements, population dynamics, and ecosystem processes. Pages 296–340 *in* C. J. Zabel and R. G. Anthony, editors. Mammal community dynamics: management and conservation in the coniferous forests of western North America. Cambridge University Press, New York, New York, USA.

King, M. M., and H. D. Smith. 1980. Differential habitat utilization by the sexes of mule deer. Great Basin Naturalist 40:273–281.

Kioko, J., A. Horton, M. Libre, J. Vickers, E. Dressel, H. Kasey, P. M. Ndegeya, D. Gadiye, B. Kissui, and C. Kiffner. 2020. Distribution and abundance of African elephants in Ngorongoro Crater, northern Tanzania. African Zoology 55:303–310.

Kirby, T., G. Shannon, B. Page, and R. Slotow. 2008. The influence of sexual dimorphism on the foraging behaviour of the nyala *Tragelaphus angasii*. Acta Zoologica Sinica 54:561–568.

Kitchen, D. W. 1974. Social behavior and ecology of the pronghorn. Wildlife Monographs 38:1–96.

Kittle, A. M., M. Anderson, T. Avgar, J. A. Baker, G. S. Brown, J. Hagens, E. Iwachewski, S. Moffatt, A. Mosser, B. R. Patterson, et al. 2015. Wolves adapt territory size, not pack size to local habitat quality. Journal of Animal Ecology 84:1177–1186.

Kleiman, D. G., and J. R. Malcom. 1981. The evolution of male parental investment in mammals. Pages 347–387 *in* D. J. Gubernick and P. H. Klopfer, editors. Parental care in mammals. Plenum Press, New York, New York, USA.

Klein, D. R. 1968. The introduction, increase, and crash of reindeer on St. Mathew Island. Journal of Wildlife Management 32:350–367.

Klich, D., and M. R. Magomedov. 2010. Abundance, population structure and seasonally changing social organization of argali *Ovis ammon karelini* in West-Central Tian-Shan of Kyrgyzstan. Acta Theriologica 55:27–34.

Klimley, A. P. 1987. The determinants of sexual segregation in the scalloped hammerhead shark, *Sphyrna lewini*. Environmental Biology of Fishes 18:2740.

Kock, A., M. J. O'Rian, K. Mauff, M. Meyer, D. Kotze, and C. Griffiths. 2013. Residency, habitat use and sexual segregation of white sharks, *Carcharodon carcharias*, in False Bay, South Africa. PLoS ONE 8:e55048.

Kodric-Brown, A., R. M. Sibly, and J. H. Brown. 2006. The allometry of ornaments and weapons. Proceedings of the National Academy of the United States of America 103:8733–8738.

Koga, T., and Y. Ono. 1994. Sexual differences in foraging behavior of sika deer, *Cervus nippon*. Journal of Mammalogy 75:129–135.

Kohl, M. T., T. K. Ruth, D. T. Metz, D. W. Smith, P. J. White, and D. R. MacNulty. 2019. Do prey select for vacant hunting domains to minimize a multi-predator threat? Ecology Letters 22:1724–1733.

Komar, O., B. J. O'Shea, A. T. Peterson, and A. G. Navarro-Sigüenza. 2005. Evidence of latitudinal sexual segregation among migratory birds wintering in Mexico. Auk 122:938–948.

Komers, P. E. 1996. Obligate monogamy without paternal care in Kirk's dikdik. Animal Behaviour 51:131–140.

Komers, P. E., F. Messier, and C. C. Gates. 1994. Plasticity of reproductive behaviour in wood bison bulls: when subadults are given a chance. Ethology Ecology and Evolution 6:313–330.

Kopecký, O., J. Vojar, F. Šusta, and I. Rehák. 2012. Composition and scaling of male and female alpine newt (*Mesotriton alpestris*) prey with related site and seasonal effects. Annales Zoologici Fennici 49:231–239.

Korpelainen, H. 1991. Sex-ratio variation and spatial segregation of the sexes in populations of *Rumex acetosa* and *R. acetosella* (Polygonaceae). Plant Systematics and Evolution 174:183–195.

Koutnik, D. L. 1981. Sex-related differences in the seasonality of agonistic behavior in mule deer. Journal of Mammalogy 62:1–11.

Kovacs, K. M., K. M. Jonas, and S. E. Welke. 1990. Sex and age segregation by *Phoca vitulina concolor* at haul-out sites during the breeding season in the Passamaquoddy Bay region, New Brunswick. Marine Mammal Science 6:204–214.

Kowalczyk, R., J. M. Wójcik, P. Taberlet, T. Kamiński, C. Miquel, A. Valentini, J. M. Craine, and E. Coissac. 2019. Foraging plasticity in a sheltering forest habitat: DNA metabarcoding diet analysis of the European bison. Forest Ecology and Management 449:117474.

Krasinska, M., and Z. A. Krasinski. 1995. Composition, group size, and spatial distribution of European bison bulls in Bialowieza Forest. Acta Theriologica 28:273–299.

Krause, J., and G. D. Ruxton. 2002. Living in groups. Oxford Series in Ecology and Evolution. Oxford University Press, Oxford, United Kingdom.

Krausman, P. R. 2002. Introduction to wildlife management: the basics. Prentice Hall, Upper Saddle River, New Jersey, USA.

Krausman, P. R., and V. C. Bleich. 2013. Conservation and management of ungulates in North America. International Journal of Environmental Studies 70:372–382.

Krausman, P. R., and J. W. Cain III, editors. 2013. Wildlife management and conservation: contemporary principles and practices. Johns Hopkins University Press, Baltimore, Maryland, USA.

Krausman, P. K., and M. L. Morrison. 2016. Another plea for standard terminology. Journal of Wildlife Management 80:1143–1144.

Krüger, O., J. B. W. Wolf, R. M. Jonker, J. I. Hoffman, and F. Trillmich. 2014. Disentangling the contribution of sexual selection and ecology to the evolution of size dimorphism in pinnipeds. Evolution 68:1485–1496.

Kruuk, H. 1972. The spotted hyena: a study of predation and social behavior. University of Chicago Press, Chicago, Illinois, USA.

Kruuk, H. 2002. Hunter and hunted: relationship between carnivores and people. Cambridge University Press, Cambridge, United Kingdom.

Kruuk, L. E. B., J. Slate, J. M. Pemberton, S. Brotherstone, F. Guinness, and T. Clutton-Brock. 2002. Antler size in red deer: heritability and selection but no evolution. Evolution 56:1683–1695.

Kucera, T. E. 1976. Social behavior and breeding system of the desert mule deer. Journal of Mammalogy 40:420–424.

Kunkel, K. E., and D. H. Pletscher. 2001. Winter hunting patterns of wolves in and near Glacier National Park, Montana. Journal of Wildlife Management 65:520–530.

Kunz, T. H., and L. F. Lumsden. 2003. Ecology of cavity and foliage roosting bats. Pages 3–89 *in* T. H. Kunz and M. B. Fenton, editors. Bat ecology. Chicago University Press, Chicago, Illinois, USA.

Kwon, Y., A. C. Doty, M. L. Huffman, V. Rolland, D. R. Istvano, and T. S. Risch. 2019. Implications of forest management practices for sex-specific habitat use by *Nycticeius humeralis*. Journal of Mammalogy 100:1263–1273.

LaGory, K. E., C. Bagshaw III, and L. R. Brisbin Jr. 1991. Niche differences between male and female white-tailed deer on Ossabaw Island, Georgia. Applied Animal Behaviour Science 29:205–214.

Laidre, K. L., P. J. Haegerty, M. P. Jorgensen, L. Witting, and M. Simon. 2009. Sexual segregation of common minke whales (*Balaenoptera acutorostrata*) in Greenland, and the influence of sea temperature on the sex ratio of catches. Journal of Marine Science 66:2253–2266.

Lande, R. 1980. Sexual dimorphism, sexual selection, and adaptation in polygenic characters. Evolution 34:292–305.

Lande, R., S. Engen, B.-E. Sæther, and T. Coulson. 2006. Estimating density dependence from time series of population age structure. American Naturalist 168:76–87.

Langbein, J., and S. J. Thirgood. 1989. Variation in mating systems of fallow deer (*Dama dama*) in relation to ecology. Ethology 83:195–214.

Langley, I., M. Fedak. K. Nichols, and L. Boehme. 2018. Sex-related differences in the postmolt distribution of Weddell seals (*Leptonychotes weddellii*) in the southern Weddell Sea. Marine Mammal Science 34:403–419.

Lashley, M. A., M. C. Chitwood, M. T. Biggerstaff, D. L. Morina, C. E. Moorman, and D. S. DePerno. 2014. White-tailed deer vigilance: the influence of social and environmental factors. PLoS ONE 9:e90652.

Laurian, C., J.-P. Quellet, R. Courtois, L. Breton, and S. St-Onge. 2001. Effects of intensive harvesting on moose reproduction. Journal of Applied Ecology 37:515–531.

Lawrence, W. S. 1982. Sexual dimorphism in between and within patch movements of a monophagous insect: *Tetraopes* (Coleoptera: Cerambycidae). Oecologia 53:245–25.

Lazo, A. 1994. Social segregation and the maintenance of social stability in a feral cattle population. Animal Behaviour 48:1133–1141.

Leader-Williams, N. 1988. Reindeer on South Georgia: the ecology of an introduced population. Cambridge University Press, Cambridge, United Kingdom.

Lei, R., Z. Jiang, and B. Liu. 2006. Group pattern and social segregation in Przewalski's gazelle (*Procapra przewalskii*) around Qinghai Lake, China. Journal of Zoology 255:175–180.

Lema, M., N. Dehnhard, G. Luna-Jorquera, C. C. Voit, and S. Garthe. 2020. Breeding stage, not sex, affects foraging characteristics in masked boobies at Rapa Nui. Behavioral Ecology and Sociobiology 74:149.

Lenart, E. A., R. T. Bowyer, J. Ver Hoef, and R. W. Ruess. 2002. Climate change and caribou: effects of summer weather on forage. Canadian Journal of Zoology 80:664–678.

Lenarz, M. S. 1985. Lack of diet segregation between sexes and age groups in feral horses. Canadian Journal of Zoology 63:2583–2585.

Lendrum, P. E., C. R. Anderson Jr., R. L. Long, J. G. Kie, and R. T. Bowyer. 2012. Habitat selection by mule deer during migration: effects of landscape structure and natural-gas development. Ecosphere 3:82.

Lendrum, P. E., C. R. Anderson Jr., K. L. Monteith, J. A. Jenks, and R. T. Bowyer. 2013. Migrating mule deer: effects of anthropogenically altered landscapes. PLoS ONE 8:e64548.

Lendrum, P. E., C. R. Anderson Jr., K. L. Monteith, J. A. Jenks, and R. T. Bowyer. 2014. Relating the movement of a rapidly migrating ungulate to spatiotemporal patterns of forage quality. Mammalian Biology 79:369–375.

Lent, P. C. 1965. Rutting behaviour in barren-ground caribou population. Animal Behaviour 13: 259–264.

Lent, P. C. 1974. Mother-infant relationships in ungulates. Pages 14–53 *in* V. Geist and F. Walther, editors. The behaviour of ungulates and its relation to management. New Series 24. International Union for the Conservation of Nature and Natural Resources, Morges, Switzerland.

Leopold, A. 1933. Game management. Charles Scribner's Sons, New York, New York, USA.

Leopold, A. 1943. Deer irruptions. Wisconsin Conservation Bulletin 8(3):3–11.

Leopold, A. 1966. A Sand County almanac. Oxford University Press, New York, New York, USA.

Le Pendu, Y., C. Guilhem, L. Briedermann, M.-L. Maublanc, and J.-F. Gerard. 2000. Interactions and associations between age and sex classes in mouflon sheep (*Ovis gmelini*) during winter. Behavioural Processes 52:97–107.

Lesage, L., M. Créte, J. Huot, and J.-P. Ouellet. 2002. Use of forest maps versus field surveys to measure summer habitat selection and sexual segregation in northern white-tailed deer. Canadian Journal of Zoology 80:717–726.

Leslie, D. M., Jr., and C. L. Douglas. 1979. Desert bighorn sheep of the River Mountains, Nevada. Wildlife Monographs 66:1–56.

Leslie, D. M., Jr., R. T. Bowyer, and J.A. Jenks. 2008. Facts from feces: nitrogen still measures up as a nutritional index for mammalian herbivores. Journal of Wildlife Management 72:1420–1433.

Leslie, D. M., Jr., R. B. Soper, R. L. Lochmiller, and D. M. Engle. 1996. Habitat use by white-tailed deer on Cross Timbers rangeland following brush management. Journal of Range Management 49:401–406.

Leuthold, W. 1977. African ungulates: a comparative review of their ethology and behavioral ecology. Springer-Verlag, Berlin, Germany.

Levin, E., U. Roll, A. Dolev, Y. Yom-Tov, and N. Kronfeld-Shcor. 2013. Bats of a gender flock together: sexual segregation in a subtropical bat. PLoS ONE 8:54987.

Lewis, R., T. O'Connell, M. Lewis, C. Campagna, and A. Hoelzel. 2006. Sex-specific foraging strategies and resource partitioning in the southern elephant seal (*Mirounga leonina*). Proceedings of the Royal Society of London B 273:2901–2907.

Li, B., M. Wang, D. A. Blank, W. Xu, W. Yang, and K. E. Ruckstuhl. 2017. Sexual segregation in the Darwin's wild sheep, *Ovis ammon darwini*, (Bovidae, Artiodactyla), in the Mengluoke Mountains of Xinjiang, China. Folia Zoologica 66:126–132.

Li, C., Z. Jiang, L. Li, Z. Li, H. Fang, C. Li, and G. Beauchamp. 2012. Effects of reproductive status, social rank, sex and group size on vigilance patterns in Przewalski's gazelle. PLoS ONE 7:e32607.

Li, Z., Z. Jiang, and G. Beauchamp. 2009. Vigilance in Przewalski's gazelle: effects of sex, predation risk and group size. Journal of Zoology 277:302–308.

Lincoln, G. A. 1971. Seasonal reproductive changes in the red deer stag (*Cervus elaphus*). Journal of Zoology 165:105–123.

Lindenfors, P., B. S. Tullbert, and M. Biuw. 2002. Phylogenetic analyses of sexual selection and sexual size dimorphism in pinnipeds. Behavioral Ecology and Sociobiology 52:188–193.

Lingle, S. 2000. Seasonal variation in coyote feeding behaviour and mortality of white-tailed deer and mule deer. Canadian Journal of Zoology 78:85–99.

Linnell, J. D. C., and F. E. Zachos. 2011. Status and distribution patterns of European ungulates: genetics, population history and conservation. Pages 12–53 *in* R. Putman, M. Apollonio, and R. Andersen, editors. Ungulate management in Europe: problems and practices. Cambridge University Press, Cambridge, United Kingdom.

Linton, D., and D. W. Macdonald. 2019. Roost composition and sexual segregation in a lowland population of

Daubenton's bats (*Myotis daubentonii*). Acta Chiropterologica 21:129–137.

Littleford-Colquhoun, B. L., C. Clemente, G. Thompson, R. H. Cristescu, N. Peterson, K. Strickland, D. Stuart-Fox, and C. H. Frere. 2019. How sexual and natural selection shape sexual size dimorphism: evidence from multiple evolutionary scales. Functional Ecology 33:1446–1458.

Litvaitis, J. A. 1990. Differential habitat use by sexes of snowshoe hares (*Lepus americanus*). Journal of Mammalogy 71:520–523.

Lodberg-Holm, H. K., S. M. J. G. Steyaert, S. Reinhardt, and F. Rosell. 2021. Size is not everything: differing activity and foraging patterns between the sexes in a monomorphic mammal. Behavioral Ecology and Sociobiology 75:76.

Lodé, T. 1996. Conspecific tolerance and sexual segregation in the use of space and habitats in the European polecat. Acta Theriologica 41:71 176.

Loe, L. E., R. J. Irving, C. Bonenfant, A. Stien, R. Langvatn, S. D. Albon A. Mysterud, and N. C. Stenseth. 2006. Testing five hypotheses of sexual segregation in an arctic ungulate. Journal of Animal Ecology 75:485 496.

Loe, L. E., G. E. Liston, G. Pigeon, K. Barker, N. Horvitz, A. Stien, M. Forchhammer, W. M. Getz, R. J. Irvine, A. Lee, et al. 2021. The neglected season: warmer autumns counteract harsher winters and promote population growth in Arctic reindeer. Global Change Biology 27.993-1002.

Logan, K. A., and L. L. Sweanor. 2001. Desert puma: evolutionary ecology and conservation of an enduring carnivore. Island Press, Washington, DC, USA.

Loison, A., J.-M. Gaillard, C. Pelabon, and N. G. Yoccoz. 1999. What factors shape sexual size dimorphism in ungulates? Evolutionary Ecology Research 1:611–633.

Long, R. A., R. T. Bowyer, W. P. Porter, P. Mathewson, K. L. Monteith, S. L. Findholt, B. L. Dick, and J. G. Kie. 2016. Linking habitat selection to fitness-related traits in herbivores: the role of the energy landscape. Oecologia 181:709–720.

Long, R. A., R. T. Bowyer, W. P. Porter, P. Mathewson, K. L. Monteith, and J. G. Kie. 2014. Behavior and nutritional condition buffer a large-bodied endotherm against direct and indirect effects of climate. Ecological Monographs 84:513–532.

Long, R. A., J. G. Kie, R. T. Bowyer, and M. A. Hurley. 2009a. Resource selection and movements by female mule deer *Odocoileus hemionus*: effects of reproductive stage. Wildlife Biology 15:288–298.

Long, R. A., J. L. Rachlow, and J. G. Kie. 2009b. Sex-specific responses of North American elk to habitat manipulation. Journal of Mammalogy 90:423–432.

Loseto, L. L., P. Richard, G. A. Stern, J. Orr, and S. H. Ferguson. 2006. Segregation of Beaufort Sea beluga whales during the open-water season. Canadian Journal of Zoology 84:1743–1751.

Lotka, A. J. 1925. Elements of physical biology. Williams and Wilkins, Baltimore, Maryland, USA.

Lovari, S., and S. B. Ale. 2001. Are there multiple mating strategies in blue sheep? Behavioural Processes 53:131–135.

Ludynia, K., N. Dehnhard, M. Poisbleau, J. F. Masello, C. C. Voight, and P. Quillfeldt. 2013. Sexual segregation in rockhopper penguins during incubation. Animal Behaviour 85:255–267.

Lukas, D., and T. H. Clutton-Brock. 2013. The evolution of social monogamy in mammals. Science 341:526–530.

Lundrigan, B. 1996. Morphology of horns and fighting behavior in the family Bovidae. Journal of Mammalogy 77:462–475.

MacArthur, R. H. 1968. The theory of niche. Pages 159–176 in R. C. Lewontin, editor. Population biology and evolution. Syracuse University Press, Syracuse, New York, USA.

MacCracken, J. G., and V. Van Ballenberghe. 1987. Age- and sex-related differences in fecal pellet dimensions of moose. Journal of Wildlife Management 51:360–364.

Macdonald, D. W. 1983. The ecology of carnivore social behaviour. Nature 301:379–384.

Macdonald, D. W., and C. Sillero-Zubiri. 2004. Wild canids—an introduction and *dramatis personae*. Pages 3 36 in D. W. Macdonald and C. Sillero-Zubiri, editors. The biology and conservation of wild canids. Oxford University Press, Oxford, United Kingdom.

MacFarlane, A. M. 2006. Can the activity budget hypothesis explain sexual segregation in western grey kangaroos? Behaviour 143:1123–1143.

MacFarlane, A. M., and G. Coulson. 2005. Sexual segregation in Australian marsupials. Pages 254–279 in K. E. Ruckstuhl and P. Neuhaus, editors. Sexual segregation in vertebrates: ecology of the two sexes. Cambridge University Press, Cambridge, United Kingdom.

MacFarlane, A. M., and G. Coulson. 2007. Sexual segregation in western grey kangaroos: testing alternative evolution hypotheses. Journal of Zoology 273:220–228.

MacFarlane, A. M., and G. Coulson. 2009. Boys will be boys: social affinity among males drives social segregation in western grey kangaroos. Journal of Zoology 277:37–44.

Mackie, R. J., K. L. Hamlin, D. F. Pac, G. L. Dusek, and A. K. Wood. 1990. Compensation in free-ranging deer populations. Transactions of the North American Wildlife and Natural Resources Conference 55:518–526.

Mac Nally, R., R. P. Duncan, J. R. Thompson, and J. D. L. Yen. 2018. Model selection using information criteria, but is the "best" model any good? Journal of Applied Ecology 55:1441–444.

Maehr, D. S. 1997. The Florida panther: life and death of a vanishing carnivore. Island Press, Washington, DC, USA.

Mahoney, S. P., and V. Geist, editors. 2019. The North American model of wildlife conservation. Johns Hopkins University Press, Baltimore, Maryland, USA.

Maier, J. A. K., and E. Post. 2001. Sex-specific dynamics of North American elk in relation to global climate. Alces 37:411–429.

Maier, J. A. K., J. M. Ver Hoef, A. D. McGuire, R. T. Bowyer, L. Saperstein, and H. A. Maier. 2005. Distribution and density of moose in relation to landscape characteristics: effects of scale. Canadian Journal of Forest Research 35:2233–2243.

Main, M. B. 1998. Sexual segregation in ungulates: a reply. Journal of Mammalogy 79, 1410–1415.

Main, M. B. 2008. Reconciling competing ecological explanations for sexual segregation in ungulates. Ecology 89:693–704.

Main, M. B., and B. E. Coblentz. 1990. Sexual segregation among ungulates: a critique. Wildlife Society Bulletin 48:2014–210.

Main, M. B., and B. E. Coblentz. 1996. Sexual segregation in Rocky Mountain mule deer. Journal of Wildlife Management 60:497–507.

Main, M. B., and J. T. du Toit. 2005. Sex differences in reproductive strategies affect habitat choice in ungulates. Pages 148–161 in K. E. Ruckstuhl and P. Neuhaus, editors. Sexual segregation in vertebrates: ecology of the two sexes. Cambridge University Press, Cambridge, United Kingdom.

Main, M. B., F. W. Weckerly, and V. C. Bleich. 1996. Sexual segregation in ungulates: new directions for research. Journal of Mammalogy 77:449–461.

Malagnino, A., P. Marchand, M. Garel, B. Cargnelutti, C. Itty, Y. Chyaval, A. J. M. Hewsion, A. Loison, and N. Morellet. 2021. Do reproductive constraints or experience drive age-dependent space use in two large herbivores? Animal Behaviour 172:121–133.

Mallory, C. D., and M. S. Boyce. 2018. Observed and predicted effects of climate change on arctic caribou and reindeer. Environmental Reviews 26:13–25.

Manly, B. F., L. McDonald, D. Thomas, T. L. McDonald, and W. P. Erickson. 2002. Resource selection by animals: statistical design and analysis for field studies. Second edition. Springer, Dordrecht, Germany.

Mao, J. S., M. S. Boyce, D. W. Smith, F. J. Singer, D. J. Vales, J. M. Vore, and E. H. Merrill. 2010. Habitat selection by elk before and after wolf reintroduction in Yellowstone National Park. Journal of Wildlife Management 69:1691–1701.

Marcelli, M., R. Fusillo, and L. Boitani. 2006. Sexual segregation in the activity patterns of European polecats (Mustela putorius). Journal of Zoology 261:249–255.

Marchand, P. N., M. Garel, G. Bourgoin, D. Dubray, D. Maillard, and A. Loison. 2015. Coupling scale-specific habitat selection and activity reveals sex-specific food/cover trade-offs in a large herbivore. Animal Behaviour 102:169–187.

Marchinton, R. L., and D. H. Hirth. 1984. Behavior. Pages 129–168 in L. K. Halls, editor. White-tailed deer: ecology and management. Stackpole Books, Harrisburg, Pennsylvania, USA.

Mariano-Jelicich, E. Madrid, and M. Favero. 2007. Sexual dimorphism and diet segregation in the black skimmer Rynchops niger. Ardea 95:115–124.

Markussen, S. S., I. Herfindall, A. Loison, E. J. Solberg, H. Haanes, K. H., Røed, M. Hein, and B.-E. Sæther. 2019. Determinations of age at first reproduction and life-time breeding success revealed by full paternity assignment in a male ungulate. Oikos 128:328–337.

Marlowe, F. W. 2000. Paternal investment and the human mating system. Behavioural Processes 51:45–61.

Marra, P. P., and R. T. Holmes. 2001. Consequences of dominance-mediated habitat segregation in American redstarts during the nonbreeding season. Auk 118:92–104.

Marshal, J. P., P. R. Krausman, and V. C. Bleich. 2005. Dynamics of mule deer forage in the Sonoran Desert. Journal of Arid Environments 60:593–609.

Martinez-Bakker, M., K. M. Bakker, and A. A. King. 2014. Human birth seasonality: latitudinal gradient and interplay with childhood disease dynamics. Proceedings of the Royal Society B 281:20132438.

Mautz, W. W. 1978. Sledding on a brushy hillside: the fat cycle in deer. Wildlife Society Bulletin 6:88–90.

Maynard Smith, J. 1965. The evolution of alarm calls. American Naturalist 99:59–63.

McCabe, R. E. 2002. Elk and Indians: then again. Pages 121–197 in D. E. Toweill and J. W. Thomas, editors. North American elk: ecology and management. Smithsonian Institution Press, Washington, DC, USA.

McCauley, D. J., F. Keesing, T. P. Young, B. F. Allan, and R. M. Pringle. 2006. Indirect effects of large herbivores on snakes in an African Savanna. Ecology 87:2657–2663.

McCulley, A. M., K. L. Parker, and M. P. Gillingham. 2017. Yukon moose: I. Seasonal resource selection by males and females in a multi-predator boreal ecosystem. Alces 53:113–136.

McCullough, D. R. 1979. The George Reserve deer herd: ecology of a K-selected species. University of Michigan Press, Ann Arbor, Michigan, USA.

McCullough, D. R. 1984. Lessons from the George Reserve, Michigan. Pages 211–242 in L. K. Halls, editor. White-tailed deer ecology and management. Stackpole Books, Harrisburg, Pennsylvania, USA.

McCullough, D. R. 1990. Detecting density dependence: filtering the baby from the bathwater. Transactions of the North American Wildlife and Natural Resources Conference 55:534–543.

McCullough, D. R. 1997. Irruptive behavior in ungulates. Pages 69–98 in W. J. McShea, H. B. Underwood, and J. H. Rappole, editors. The science of overabundance: deer ecology and population management. Smithsonian Institution Press, Washington, DC, USA.

McCullough, D. R. 1999. Density dependence and life-history strategies of ungulates. Journal of Mammalogy 80:1130–1146.

McCullough, D. R. 2001. Male harvest in relation to female removals in a black tailed deer population. Journal of Wildlife Management 65:46–58.

McCullough, D. R., and Y. McCullough. 2000. Kangaroos in outback Australia: comparative ecology and behavior of three coexisting species. Columbia University Press, New York, New York, USA.

McCullough, D. R., D. H. Hirth, and S. J. Newhouse. 1989. Resource partitioning between sexes in white-tailed deer. Journal of Wildlife Management 53:277–283.

McHugh, T. 1972. The time of the buffalo. Alfred A. Knopf, New York, New York, USA.

McIntyre, C. Tosh, J. Plötz, H. Bornemann, and M. Bester. 2010. Segregation in a sexually dimorphic mammal: a mixed-effects modeling analysis of diving behaviour in southern elephant seals. Marine Ecology Progress Series 412:293–304.

McKee, J., E. Chambers, and J. Guseman. 2013. Human population density and growth validated as extinction threats to mammal and bird species. Human Ecology 41:773–778.

McNab, B. K. 1963. Bioenergetics and determination of home range size. American Naturalist 97:133–140.

McNaughton, S. J. 1979a. Grazing as an optimization process: grass-ungulate relationships in the Serengeti. American Naturalist 113:691–703.

McNaughton, S. J. 1979b. Grassland-herbivore dynamics. Pages 46–81 in A. R. E. Sinclair and M. Norton-Griffiths, editors. Serengeti: dynamics of an ecosystem. University of Chicago Press, Chicago, Illinois, USA.

McNitt, D. C., R. S. Alonso, M. J. Cherry, M. L. Fies, and M. J. Kelly. 2020. Sex-specific effects of reproductive season on bobcat space use, movement, and resource selection in the Appalachian Mountains of Virginia. PLoS ONE 15:e0225355.

McShea, W. J., M. Aung, D. Poszig, C. Wemmer, and S. Monfort. 2001. Forage, habitat use, sexual segregation by a tropical deer (Cervus eldi thamin) in a dipterocarp forest. Journal of Mammalogy 82:848–857.

Mech, L. D. 1970. The wolf: the ecology and behavior of an endangered species. Natural History Press, Garden City, New York, USA.

Mech, L. D. 2012. Is science in danger of sanctifying the wolf? Biological Conservation 150:143–149.

Mech, L. D., and R. O. Peterson. 2003. Wolf-prey relations. Pages 131–160 in L. D. Mech and L. Boitani, editors. Wolves: behavior, ecology and conservation. University of Chicago Press, Chicago, Illinois, USA.

Mech, L. D., R. J. Meier, J. W. Burch, and L. G. Adams. 1995. Patterns of prey selection by wolves in Denali National Park, Alaska. Pages 231–244 in L. N. Carbyn, S. H. Fritts, and D. R. Seip, editors. Ecology and conservation of wolves in a changing world. Canadian Circumpolar Institute, Occasional Publication 35, University of Alberta, Edmonton, Alberta, Canada.

Meldrum, G. E., and K. E. Ruckstuhl. 2009. Mixed-sex group formation by bighorn sheep in winter: trading costs of synchrony for benefits of group living. Animal Behaviour 77:919–929.

Mercer, C. A., and S. M. Epplex. 2010. Inter sexual competition in a dioecious grass. Oecologia 164:657–664.

Merems, J. L., L. A. Shipley, T. Levi, J. Ruprecht, D. A. Clark, M. J. Wisdom, N. J. Jackson, K. M. Stewart, and R. A. Long. 2020. Nutritional-landscape models link habitat use to condition of mule deer (Odocoileus hemionus). Frontiers in Ecology and Evolution 8:98.

Merrill, E., J. Killeen, J. Pettit, M. Trottier, H. Martin, J. Berg, H. Bohm, S. Eggeman, and M. Hebblewhite. 2020. Density-dependent foraging behaviors on sympatric wither ranges in a partially migratory elk population. Frontiers in Ecology and Evolution 8:269.

Michaud, R. 2005. Sociality and ecology of odontocetes. Pages 303–326 in K. E. Ruckstuhl and P. Neuhaus, editors. Sexual segregation in vertebrates: ecology of the two sexes. Cambridge University Press, Cambridge, United Kingdom.

Michelena, P., P. M. Bouquet, A. Gissac, V. Fourcassie, J. Lauga, J.-F. Gerard, and R. Bon. 2004. An experimental test of hypotheses explaining social segregation in dimorphic ungulates. Animal Behaviour 68:1371–1380.

Michelena, P., J. Gautrais, J.-F. Gérard, R. Bon, and J. L. Deneubourg. 2008. Social cohesion in groups of sheep: effect of activity level, sex composition and group size. Applied Animal Behaviour Science 112:81–93.

Michelena, P., S. Noël, J. Gautrais, J.-F. Gerard, J.-L. Deneubourg, and R. Bon. 2006. Sexual dimorphism, activity budget and synchrony in groups of sheep. Oecologia 148:170–180.

Mielke, P. W., and K. J. Berry. 2001. Permutation methods: a distance function approach. Springer Series in Statistics. Springer, New York, New York, USA.

Mielke, P. W., J. C. Anderson, and K. J. Berry. 1983. Lead concentrations in inner-city soils as a factor in the child lead problem. American Journal of Public Health 73:1366–1369.

Milankovic, H. R., N. D. Ray, L. K. Gentle, C. Kruger, E. Jacobs, and C. J. Ferreira. 2021. Seasonal occurrence and sexual segregation of great white sharks *Carcharodon carcharias* in Mossel Bay, South Africa. Environmental Biology of Fishes 104:555–568.

Millar, J. S., and R. M. Zammuto. 1983. Life histories of mammals: an analysis of life tables. Ecology 64:631–635.

Miller, B. K., and J. A. Litvaitis. 1992. Habitat segregation by moose in a boreal forest ecotone. Acta Theriologica 37:41–50.

Miller, M. G. R., F. R. O. Silva, G. E. Machovsky-Capuska, and B. C. Congdon. 2018. Sexual segregation in tropical seabirds: drivers of sex-specific foraging in the brown booby *Sula leucogaster*. Journal of Ornithology 159:425–437.

Miquelle, D. G. 1990. Why don't bull moose eat during the rut? Behavioral Ecology and Sociobiology 27:145–151.

Miquelle, D. G., J. M. Peek, and V. Van Ballenberghe. 1992. Sexual segregation in Alaskan moose. Wildlife Monographs 122:1–57.

Miranda, M., M. Sicilia, J. Bartolomé, E. Molina-Alcaide, L. Gálvez-Bravo, and J. Cassinello. 2012. Foraging sexual segregation in a Mediterranean environment: summer drought modulates sex-specific resource selection. Journal of Arid Environments 85:97–104.

Mirarchi, R. E., B. E. Howland, P. F. Scanlon, and R. L. Kirkpatrick. 1978. Seasonal variation in plasma LH, FSH, prolactin, and testosterone in plasma in adult male white-tailed deer. Canadian Journal of Zoology 56:121–127.

Moehlman, P. D. 1986. Ecology of cooperation in canids. Pages 64–86 *in* D. I. Rubenstein and R. W. Wrangham, editors. Ecology aspects of social evolution. Princeton University Press, Princeton, New Jersey, USA.

Molvar, E. M., and R. T. Bowyer. 1994. Costs and benefits of group living in a recently social ungulate: the Alaskan moose. Journal of Mammalogy 75: 621–630.

Molvar, E. M., R. T. Bowyer, and V. Van Ballenberghe. 1993. Moose herbivory, browse quality, and nutrient cycling in an Alaskan treeline community. Oecologia 94: 472–479.

Moncorps, S. Boussès, D. Réale, and J.-L. Chapuis. 1997. Diurnal time budget of the mouflon (*Ovis musimon*) on the Kerguelen archipelago: influence of food resources, age, and sex. Canadian Journal of Zoology 75:1828–1834.

Monteith, K. B., K. L. Monteith, R. T. Bowyer, D. M. Leslie Jr., and J. A. Jenks. 2014a. Reproductive effects on fecal nitrogen as an index of diet quality: an experimental assessment. Journal of Mammalogy 95:301–310.

Monteith, K. L., V. C. Bleich, T. R. Stephenson, B. M. Pierce, M. M. Conner, J. G. Kie, and R. T. Bowyer. 2014b. Life-history characteristics of mule deer: effects of nutrition in a variable environment. Wildlife Monographs 186:1–56.

Monteith, K. L., V. C. Bleich, T. R. Stephenson, B. M. Pierce, M. M. Conner, R. W. Klaver, and R. T. Bowyer. 2011. Timing of seasonal migration in mule deer: effects of climate, plant phenology, and life-history characteristics. Ecosphere 2:47.

Monteith, K. L., R. W. Klaver, K. R. Hersey, A. A. Holland, T. P. Thomas, and M. J. Kauffman. 2015. Effects of climate and plant phenology on recruitment of moose at the southern extent of their range. Oecologia 178:1137–1148.

Monteith, K. L., R. A. Long, V. C. Bleich, J. R. Heffelfinger, P. R. Krausman, and R. T. Bowyer. 2013a. Effects of harvest, culture, and climate on trends in size of horn-like structures in trophy ungulates. Wildlife Monographs 183:1–26.

Monteith, K. L., R. A. Long, T. R. Stephenson, V. C. Bleich, R. T. Bowyer, and T. N. LaSharr. 2018. Horn size and nutrition in mountain sheep: can ewe handle the truth? Journal of Wildlife Management 82:67–84.

Monteith, K. L., L. E. Schmitz, J. A. Jenks, J. A. Delger, and R. T. Bowyer. 2009. Growth of male white-tailed deer: consequences of maternal effects. Journal of Mammalogy 90:651–550.

Monteith, K. L., C. L. Sexton, J. A. Jenks, and R. T. Bowyer. 2007. Evaluation of techniques for categorizing group membership of white-tailed deer. Journal of Wildlife Management 71:1712–1716.

Monteith, K. L., T. R. Stephenson, V. C. Bleich, M. M. Conner, B. M. Pierce, and R. T. Bowyer. 2013b. Risk-sensitive allocation in seasonal dynamics of fat and protein reserves in a long-lived mammal. Journal of Animal Ecology 82:377–388.

Mooring, M. S., and E. M. Rominger. 2004. Is the activity budget hypothesis the holy grail of sexual segregation? Behaviour 141:521–530.

Mooring, M. S., T. A. Fitzpatrick, J. E. Benjamin, I. C. Fraser, T. T. Nishihira, D. D. Reisig, and E. M. Rominger.

2003. Sexual segregation in desert bighorn sheep (*Ovis canadensis mexicana*). Behaviour 140:183–207.

Mooring, M. S., D. D. Reisig, E. R. Osborne, A. L. Kanallakan, B. M. Hall, E. W. Schaad, D. W. Wiseman, and R. R. Huber. 2005. Sexual segregation in bison: a test of multiple hypotheses. Behaviour 142:897–927.

Mora, C., D. P. Tittensor, S. Adl, A. G. B. Simpson, and B. Worm. 2011. How many species are there on Earth and in the ocean? PLoS Biology 9:e1001127.

Morales, M. B., J. Traba, E. Carriles, M. P. Delgado, and E. L. Garćia de la Morena. 2008. Sexual differences in microhabitat selection of breeding little bustards *Tetrax tetrax*: ecological segregation based on vegetation structure. Acta Oecologica 34:345–353.

Morano, S., K. M. Stewart, J. S. Sedinger, C. A. Nicolai, and M. Vavra. 2013. Life-history strategies of North American elk: trade-offs associated with reproduction and survival. Journal of Mammalogy 94:162–172.

Morgantini, L. E., and R. J. Hudson. 1981. Sex differential in use of the physical environment by bighorn sheep (*Ovis canadensis*). Canadian Field-Naturalist 95:69–74.

Morris, D. W. 1984. Sexual differences in habitat use by small mammals: evolutionary strategy or reproductive constraint? Oecologia 65:51–57.

Morris D. W., and J. T. MacEachern. 2010. Sexual conflict over habitat selection: the game and a test with small mammals. Evolutionary Ecology Research 12:507–522.

Morse, D. H. 1968. A quantitative study of foraging of male and female spruce-woods warblers. Ecology 49:779–784.

Morteo, E., A. Rocha-Olivares, and L. G. Abarca-Arenas. 2014. Sexual segregation in coastal bottlenose dolphins (*Tursiops truncates*) in the southwestern Gulf of Mexico. Aquatic Mammals 40:375–385.

Morton, T. L., J. W. Haefner, V. Nugala, R. D. Decino, and L. Mendes. 1994. The selfish herd revisited: do simple movement rules reduce relative predation risk? Journal of Theoretical Biology 167:73–79.

Mramba, R. P., O. Mahenya, A. Siyaya, K. M. Matthisen, H. P. Andreassen, and C. Skarpe. 2017. Sexual segregation in foraging giraffe. Acta Oecologica 79:26–35.

Mucientes, G. R., N. Queiroz, L. L. Souza, P. Tarroso, and D. W. Sims. 2009. Sexual segregation in sharks and the potential threat from fisheries. Biology Letters 5:156–159.

Murie, O. J. 1935. Alaska-Yukon caribou. North American Fauna 54:1–93.

Myers, P. 1978. Sexual dimorphism in size of vespertilionid bats. American Naturalist, 112:701–711.

Mysterud, A. 2000. The relationship between ecological segregation and sexual body size dimorphism in large herbivores. Oecologia 124:40–54.

Mysterud, A., T. Coulson, and N. C. Stenseth. 2002. The role of males in the dynamics of ungulate populations. Journal of Animal Ecology 71: 907–915.

Mysterud, A., F. J. Pérez-Barbería, and I. J. Gordon. 2001. The effect of season, sex, and feeding style on home range areas versus body mass scaling in temperate ruminants. Oecologia 127:30–39.

Nakashima, Y., and Y. Hirose. 2003. Sex differences in foraging behaviour and oviposition site preference in an insect predator, *Orius sauteri*. Entomologia Experimentalis et Applicata 106:79–86.

Nali, R. C., K. R. Zamudio, C. F. B. Haddad, and C. P. A. Prado. 2014. Size-dependent selective mechanisms on males and females and the evolution of sexual size dimorphism in frogs. American Naturalist 184:727–740.

Nardone, V., L. Cistrone, I. Di Salvo, A. Ariano, A. Migliozzi, C. Allgrini, L. Ancillotto, A. Fulco, and D. Russo. 2015. How to be a male at different elevations: ecology of intra-sexual segregation in the trawling bat *Myotis daubentoniid*. PLoS ONE 10:30134573

Neuhaus, P., and K. E. Ruckstuhl. 2002a. Foraging behaviour in alpine ibex (*Capra ibex*): consequences of reproductive status, body size, age, and sex. Ecology Ethology and Evolution 14:373–81.

Neuhaus, P., and K. E. Ruckstuhl. 2002b. The link between sexual dimorphism, activity budgets, and group cohesion: the case of the plains zebra (*Equus burchelli*). Canadian Journal of Zoology 80:1437–1441.

Neuhaus, P., and K. E. Ruckstuhl. 2004. A critique: can the activity budget hypothesis explain sexual segregation in desert bighorn sheep? Behaviour 141:513–520.

Neuhaus, P., K. E. Ruckstuhl, and L. Conradt. 2005. Conclusions and future directions. Pages 395–402 *in* K. E. Ruckstuhl and P. Neuhaus, editors. Sexual segregation in vertebrates: ecology of the two sexes. Cambridge University Press, Cambridge, United Kingdom.

Nicholson, M.C., R. T. Bowyer, and J. G. Kie. 1997. Habitat selection and survival of mule deer: tradeoffs associated with migration. Journal of Mammalogy 78:483–504.

Nixon, C. M., L. P. Hansen, P. A. Brewer, and J. E. Chelsvig. 1991. Ecology of white-tailed deer in an intensively farmed region of Illinois. Wildlife Monographs 118:1–77.

Noyes, J. H., B. K. Johnson, L. D. Bryant, S. L. Findholt, and J. W. Thomas. 1996. Effects of bull age on conception dates and pregnancy rates of cow elk. Journal of Wildlife Management 60:508–517.

Oates, B. A., K. L. Monteith, J. R. Goheen, J. A. Merkle, G. L. Fralick, and M. J. Kauffman. 2020. Detecting resource limitation in a large herbivore population is enhanced with measure of nutritional condition. Frontiers in Ecology and Evolution 8:522174.

Oakes, E. I., R. Harmsen, and C. Eberl. 1992. Sex, age, and seasonal differences in the diets and activity budgets of muskoxen (*Ovibos moschatus*). Canadian Journal of Zoology 70:605–616.

O'Brien, P. O., Q. M. R. Webber, and E. Vander Wal. 2018. Consistent individual differences and population plasticity in network-derived sociality: an experimental manipulation of density in a gregarious ungulate. PLoS ONE 13:e0193425.

Odum, E. P. 1959. Fundamentals of ecology. Second edition. Saunders, Philadelphia, Pennsylvania, USA.

Oehlers, S. A., R. T. Bowyer, F. Huettmann, D. K. Person, and W. B. Kessler. 2011. Sex and scale: implications for habitat selection by Alaskan moose *Alces alces gigas*. Wildlife Biology 17:67–84.

Ofstad, E. G., I. Herfindal, E. J. Solberg, M. Heirm, C. M. Rolandsen, and B.-E. Sæther. 2019. Use, selection, and home range properties: complex patterns of individual habitat utilization. Ecosphere 10:e02695.

O'Gara, B. W. 2004a. Mortality. Pages 379–408 in B. W. O'Gara and J. D. Yoakum, editors. Pronghorn: ecology and management. University Press of Colorado, Boulder, Colorado, USA.

O'Gara, B. W. 2004b. Physical characteristics. Pages 109–143 in B. W. O'Gara and J. D. Yoakum, editors. Pronghorn: ecology and management. Wildlife Management Institute, University Press of Colorado, Boulder, Colorado, USA.

Oliveira, T., F. Urra, J. M. López-Martin, E. Ballesteros-Duperón, J. M. Barea-Azcón, M. Moléon, J. M. Gil-Sánchez, P. C. Alves, F. Diaz-Ruiz, P. Ferreras, and P. Monterroso. 2018. Females know better: sex-biased habitat selection by the European wildcat. Ecology and Evolution 8:9464–9477.

Olsson, M., J. J. Cox, L. Larken, P. Widén, and A. Olovsson. 2010. Space and habitat use of moose in southwestern Sweden. European Journal of Wildlife Research 57:241–249.

Ordway, L. L., and P. R. Krausman. 1984. Habitat use by desert mule deer. Journal of Wildlife Management 50:677–683.

Organ, J. F., S. P. Mahoney, and V. Geist. 2010. Born in the hands of hunters: the North American model of wildlife conservation. The Wildlife Professional 4:22–27.

Owen-Smith, R. N. 1988. Megaherbivores: the influence of very large body size. Cambridge University Pres, Cambridge, United Kingdom.

Owen-Smith, R. N. 2006. Demographic determination of the shape of density dependence for three African ungulate populations. Ecological Monographs 76:93–109.

Ozoga, J. J., L. J. Verme, and C. S. Bienz. 1982. Parturition behavior and territoriality in white-tailed deer: impact on neonatal mortality. Journal of Wildlife Management 46:1–11.

Pagon, N., S. Grignolio, A. Pipia, P. Bongi, C. Bertolucci, and M. Apollonio. 2013. Seasonal variation of activity patterns in roe der in a temperate forested area. Chronobiology International 30:772–785.

Painter, L. E., R. L. Beschta, E. J. Larsen, and W. J. Ripple. 2018. Aspen recruitment in the Yellowstone region linked to reduced herbivory after large carnivore restoration. Ecosphere 9:e02376.

Paiva, V. H., J. A. Ramon, C. Nava, V. Neves, J. Bried, and M. Magalhães. 2018. Inter-sexual habitat and isotopic niche segregation of the endangered Monteiro's storm-petrel during breeding. Zoology 126:29–35.

Palacín, C., J. C. Alonso, J. A. Alonso, C. A. Martin, M. Magaña, and B. Martin. 2009. Differential migration by sex in the great bustard: possible consequences of an extreme sexual dimorphism. Ethology 115:617–626.

Pallante, V., D. Rucco, and E. Versace. 2021. Young chicks quickly lose their spontaneous preferences to aggregate with females. Behavioral Ecology and Sociobiology 75:78.

Pardi, M. I., and L. R. G. DeSantis. 2021. Dietary plasticity of North American herbivores: a synthesis of stable isotope data over the past 7 million years. Proceedings of the Royal Society of London B 288:20210121.

Parker, K. L., and C. T. Robbins. 1984. Thermoregulation in mule deer and elk. Canadian Journal of Zoology 62:1409–1422.

Parker, K. L., P. S. Barboza, and M. P. Gillingham. 2009. Nutrition integrates environmental responses of ungulates. Functional Ecology 23:57–69.

Pastor, J., B. Dewey, R. J. Naiman, P. F. McInnes, and Y. Cohen. 1993. Moose browsing and soil fertility of Isle Royale National Park. Ecology 74:467–480.

Payne, B. L., and J. Bro-Jørgensen. 2020. Conserving African ungulates under climate change: do communal and private conservancies fill gaps in the protected area network effectively? Frontiers in Ecology and Evolution 8:160.

Paz, J. A., J. P. Seco Pon, L. Krüger, M. Favero, and S. Copello. 2021. Is there sexual segregation in habitat selection by black-browed albatrosses wintering in the south-west Atlantic? Emu-Austral Ornithology 121:167–177.

Peek, J. M., and A. L. Lovas. 1968. Differential distribution of elk by sex and age on the Gallatin winter range, Montana. Journal of Wildlife Management 32:533–557.

Peek, J. M., R. E. LeResche, and D. R. Stevens. 1974. Dynamics of moose aggregations in Alaska, Minnesota, and Montana. Journal of Mammalogy 55:126–137.

Pellegrini, A. D., J. D. Long, and E. A. Mizerek. 2005. Sexual segregation in humans. Pages 200–217 in K. E. Ruckstuhl and P. Neuhaus, editors. Sexual segregation in vertebrates: ecology of the two sexes. Cambridge University Press, Cambridge, United Kingdom.

Pemberton, J. M., S. D. Albon, F. E. Guinness, T. H. Clutton-Brock, and G. A. Dover. 1992. Behavioral estimates of male mating success tested by DNA fingerprinting in a polygynous mammal. Behavioral Ecology 3:66–75.

Pereira, J. M., V. H. Paiva, R. A. Phillips, and J. C. Xavier. 2018. The devil is in the detail: small-scale sexual segregation despite large-scale spatial overlap in the wandering albatross. Marine Biology 165:art.55.

Pérez-Barbería, F. J, and I. J. Gordon. 1998. The influence of sexual dimorphism in body size and mouth morphology on diet selection and sexual segregation in cervids. Acta Veterinaria Hungarica 46:357–367.

Pérez-Barbería, F. J., and I. J. Gordon. 1999. Body size dimorphism and sexual segregation in polygynous ungulate: an experimental test with Soay sheep. Oecologia 120:258–267.

Pérez-Barbería, F. J., and I. J. Gordon. 2000. Differences in body mass and oral morphology between the sexes in the Artiodactyla: evolutionary relationships with sexual segregation. Evolutionary Ecology Research 2:667–684.

Pérez-Barbería, F. J., and C. Nores. 1994. Seasonal variation in group size of Cantabrian chamois in relation to escape terrain and food. Acta Theriologica 39:295–305.

Pérez-Barbería, F. J., and J. M. Yearsley. 2010. Sexual selection for fighting skills as a driver of sexual segregation in polygynous ungulates: an evolutionary model. Animal Behavior 80:745–755.

Pérez-Barbería, F. J., I. J. Gordon, and M. Pagel. 2002. The origins of sexual dimorphism in body size in ungulates. Evolution 56:1276–1285.

Pérez-Barbería, F. J., M. Olivan, K. Osoro, and C. Nores. 1997. Sex, seasonal and spatial differences in the diet of Cantabrian chamois (Rupicapra pyrenaica parva). Acta Theriologica 42:37–46.

Pérez-Barbería, F. J., E. Pérez-Fernández, E. Robertson, and B. Alvarez-Enríquez. 2008. Does the Jarman-Bell principle at intra-specific level explain sexual segregation in polygynous ungulates? Sex differences in forage digestibility in Soay sheep. Oecologia 157:21–30.

Pérez-Barbería, F. J., E. Robertson, and I. J. Gordon. 2005. Are social factors sufficient to explain sexual segregation in ungulates? Animal Behaviour 69:827–834.

Pérez-Barbería, F. J., E. Robertson, R. Soriguer, A. Aldeza-bal, M. Menizabal, and E. Pérez-Fernández. 2007. Why do polygynous ungulates segregate in space? Testing the activity-budget hypothesis in Soay sheep. Ecological Monographs 74:631–647.

Pérez-Barbería, F. J., D. M. Walker, and I. J. Gordon. 2004. Sex differences in feeding behaviour at feeding station scale in Soay sheep (Ovis aries). Behaviour 141:999–1020.

Perrig, P. A., S. A. Lambertucci, P. A. E. Alarcón, A. D. Middleton, J. Padró, P. I. Plaza, G. Blanco, J. A. Sánchez Zapata, J. Donázar, and J. N. Pauli. 2021. Limited sexual segregation in a dimorphic avian scavenger, the Andean condor. Oecologia 196:77–88.

Peterson, L. M., and F. W. Weckerly. 2017. Male group size, female distribution and changes in sexual segregation by Roosevelt elk. PLoS ONE 12:e0187829.

Peterson, R. O. 1977. Wolf ecology and prey relationships on Isle Royale. National Park Service Scientific Monograph 11:1–210.

Pettorelli, N., S. J. Ryan, T. Mueller, N. Bunnerfeld, B. Jedrzejewska, M. Lima, and K. Kausrud. 2011. The normalized difference vegetation index (NDVI) in ecology: a decade of unforeseen success. Climate. Research 46:15–27.

Phillips, D. L., R. A. R. McGill, D. A. Dawson, S. Bearhop. 2011. Sexual segregation in distribution, diet, and trophic level of seabirds: insights from stable isotope analysis. Marine Biology 158:2199–2208.

Pierce, B. M., V. C. Bleich, and R. T. Bowyer. 2000a. Selection of mule deer by mountain lions and coyotes: effects of hunting style, body size, and reproductive status. Journal of Mammalogy 81:462–472.

Pierce, B. M., V. C. Bleich, and R. T. Bowyer. 2000b. Social organization of mountain lions: does a land-tenure system regulate population size? Ecology 81:1533–1543.

Pierce, B. M., V. C. Bleich, K. L. Monteith, and R. T. Bowyer. 2012. Top-down versus bottom-up forcing: evidence from mountain lions and mule deer. Journal of Mammalogy 93:977–988.

Pilliod, D. S., C. R. Peterson, and P. I. Ritson. 2002. Seasonal migration of Columbia spotted frogs (Rana luteiventris) among complementary resources in a high mountain basin. Canadian Journal of Zoology 80:1849–1862.

Pimm, S. L., C. N. Jenkins, R. Abell, T. M. Brooks, J. L. Gittleman, L. N. Joppa, P. H. Raven, C. M. Roberts, and J. O Sexton. 2014. The biodiversity of species and their rates of extinction, distribution, and protection. Science 344:1246752.

Pipia, A., S. Ciuti, S. Grignollo, S. Luchetti, R. Madau, and M. Apollonio. 2008. Influence of sex, season, tempera-ture and reproductive status on daily activity patterns in Sardinian mouflon (Ovis orientalis musimon). Behaviour 145:1723–1745.

Pipia, A., S., Ciuti, S. Grignolio, S. Luchetti, R. Madau, and M. Apollonio. 2009. Effect of predation risk on grouping pattern and whistling behaviour in a wild mouflon *Ovis aries* population. Acta Theriologica, 54:77–86.

Pirotta, E., M. Vighi, J. M. Brotons, E. Dillane, M. Cedrà, and L. Rendell. 2020. Stable isotopes suggest fine-scale sexual segregation in an isolated, endangered sperm whale population. Marine Ecology Progress Series 654:209–218.

Plard, F., C. Monenfant, and J.-M. Gaillard. 2011. Revisiting the allometry of antlers among deer species: male-male sexual competition as a driver. Oikos 129:601–606.

Plavcan, J. M. 2012. Body size, size variation, and sexual dimorphism in early *Homo*. Current Anthropology 53:S409–S423.

Polák, J., and D. Frynta. 2009. Sexual size dimorphism in domestic goats, sheep, and their wild relatives. Biological Journal of the Linnean Society 98:872–883.

Poole, J. H. 1994. Sex differences in the behaviour of African elephants. Pages 331–346 *in* R. B. Short and E. Balaban, editors. The differences between the sexes. Cambridge University Press, Cambridge, United Kingdom.

Popper, K. R. 1959. The logic of scientific discovery. Martino Publishing, Mansfield Center, Connecticut, USA.

Post, D. M., T. S. Armbrust, E. A. Horne, and J. R. Goheen. 2001. Sexual segregation results in differences in content and quality of bison (*Bos bison*) diets. Journal of Mammalogy 82:407–413.

Post, E., and M. C. Forchhammer. 2008. Climate change reduces reproductive success of an arctic herbivore through trophic mismatch. Philosophical Transactions of the Royal Society of London B 363:2369–2375.

Powell, R. A., J. W. Zimmerman, and D. E. Seaman. 1997. Ecology and behaviour of North American black bears: home ranges, habitat and social organization. Chapman and Hall, London, United Kingdom.

Powolny, T., V. Bretagnolle, A. Dupoue, O. Lourdais, and C. Eraud. 2016. Cold tolerance and sex-dependent hypothermia may explain winter sexual segregation in a farmland bird. Physiological and Biochemical Zoology 89:151–160.

Price, S. A., O. R. P. Bininda-Emonds, and G. L. Gittleman. 2005. A complete phylogeny of the whales, dolphins and even-toed hoofed mammals (Cetartiodactyla). Biological Reviews 80:445–473.

Prieditis, A. A. 1979. Interrelation between the body weight and horn size in European elks (*Alces alces*) of different age. Zoologicheskii Zhurnal 58:105–110.

Prins, H. H. T. 1989. Condition changes and choice of social environment in African buffalo bulls. Behaviour 108:297–324.

Prins, H. H. T. 1996. Ecology and behaviour of the African buffalo: social inequality and decision making. Chapman and Hall. London, United Kingdom.

Prins, R. A., and M. J. H. Geelen. 1971. Rumen characteristics of red deer, fallow deer, and roe deer. Journal of Wildlife Management 35:673–680.

Putman, R. 1988. The natural history of deer. Cornell University Press, Ithaca, New York, New York, USA.

Putman, R., and W. T. Flueck. 2011. Intraspecific variation in biology and ecology of deer: magnitude and causation. Animal Production Science 51:277–291.

Putman, R. J., S. Culpin, and S. J. Thirgood. 1993. Dietary differences between male and female fallow deer in sympatry and allopatry. Journal of Zoology 229:267–275.

Qi, D., S. Zhang, Z. Shang, Y. Hu, X. Yang, H. Wang, and F. Wei. 2011. Different habitat preferences of male and female giant pandas. Journal of Zoology 285:205–214.

Rachlow, J. L., and J. Berger. 1997. Conservation implications of patterns of horn regeneration in dehorned white rhinos. Conservation Biology 11:84–4 91.

Rachlow, J. L., and R. T. Bowyer. 1998. Habitat selection by Dall's sheep (*Ovis dalli*): maternal trade-offs. Journal of Zoology 245:457–465.

Rachlow, J. L., E. V. Berkeley, and J. Berger. 1998. Correlates of male mating strategies in white rhinos (*Ceratotherium simum*). Journal of Mammalogy 79:1317–1324.

Ralls, K. 1977. Sexual dimorphism in mammals: avian models and unanswered questions. American Naturalist 122:917–938.

Ramesh, T., R. Kalle, and C. T. Downs. 2015. Sex-specific indicators of landscape use by servals: consequences of living in fragmented landscapes. Ecological Indicators 52:8–15.

Ramos, A., O. Pelit, P. Longour, C. Pasquaretta, and C. Sueur. 2016. Space use and movement patterns of a semi-free-ranging herd of European bison (*Bison bonasus*). PLoS ONE 11:e0147404.

Ramzinski, D. M., and F. W. Weckerly. 2007. Scaling relationship between body weight and fermentation gut capacity in axis deer. Journal of Mammalogy 88:415–420.

Ranglack, D. H., and J. T. du Toit. 2015. Habitat selection by free-ranging bison in a mixed grazing system on public land. Rangeland Ecology and Management 68:349–353.

Ransom, J. I., and P. Kaczensky, editors. 2016. Wild equids: ecology, management, and conservation. Johns Hopkins University Press, Baltimore, Maryland, USA.

Rehnus, M., and K. Bollmann. 2020. Quantification of sex-related diet composition by free-ranging mountain hares (*Lepus timidus*). Hystrix 31:80–82.

Reilly, S. M. 1983. The biology of high-altitude salamander *Batrachuperus mustersi* from Afghanistan. Journal of Herpetology 17:1–9.

Reimers, E., D. R. Klein, and R. Sørumgård. 1983. Calving time, growth rate, and body size of Norwegian reindeer on different ranges. Arctic and Alpine Research 15:107–118.

Remington, R. D., and M. A. Schork. 1970. Statistics with applications to the biological and health sciences. Prentice Hall, Englewood Cliffs, New Jersey, USA.

Reyes-González, F. De Felipe, V. Morea-Pujol, A. Soriano-Redondo, L. Navarro-Herrero, L. Zango, S. Garica-Barcelona, R. Ramos, and J. González-Solis. 2021. Sexual segregation in the foraging behaviour of a slightly dimorphic seabird: influence of the environment and fishery activity. Journal of Animal Ecology 90:1109–1121.

Reyna-Hurtago, R., E. Rojas-Flores, and G. W. Tanner. 2009. Home range and habitat preferences of white-lipped peccaries (*Tayassu pecari*) in Calakmul, Campeche, Mexico. Journal of Mammalogy 90:1199–1209

Ricca, M. A., D. H. Van Vuren, F. W. Weckerly, J. C. Williams, and A. K. Miles. 2014. Irruptive dynamics of introduced caribou on Adak Island, Alaska: an evaluation of Riney-Caughley model predictions. Ecosphere 5:94.

Richardson, K. E., and F. W. Weckerly. 2007. Intersexual social behavior of urban white-tailed deer and its evolutionary implication. Canadian Journal of Zoology 85:759–766.

Ricker, W. E. 1954 Stock and recruitment. Journal of the Fisheries Research Board of Canada 11:559–623.

Ringler, M., E. Ursprung, and W. Hodl. 2009. Site fidelity and patterns of short- and long-term movement in the brilliant-thighed poison frog *Allobates femoralis* (Aromobatidae). Behavioral Ecology and Sociobiology 63:1281–1293.

Ripple, W. J., J. A. Estes, R. L. Beschta, C. C. Wilmers, E. G. Ritchie, M. Hebblewhite, J. Berger, B. Elmhagen, M. Letnic, M. P. Nelson, et al. 2014. Status and ecological effects of the world's largest carnivores. Science 343:1241484.

Ripple, W. J., J. A. Estes, O. J. Schmitz, V. Constant, M. J. Kaylor, A. Lenz, J. L. Motley, K. E. Self, D. S. Taylor, and C. Wolf. 2016. What is a trophic cascade? Trends in Ecology and Evolution 31:824–849.

Risenhoover, K. L., and S. A. Maass. 1987. The influence of moose on the composition and structure of Isle Royale forests. Canadian Journal of Forest Research 17:357–364.

Robbins, C. T. 1983. Wildlife feeding and nutrition. Academic Press, New York, New York, USA.

Robbins, R. L. 2007. Environmental variables affecting the sexual segregation of great white sharks *Carcharodon carcharias* at the Neptune Islands of South Africa. Journal of Fish Biology 70:1350–1364.

Roberts, G. 1996. Why individual vigilance declines as group size increases. Animal Behaviour 51:1077–1086.

Robins, J. D. 1971. Differential niche utilization in a grassland sparrow. Ecology 52:1065–1070.

Robinson, R. W., T. S. Smith, J. C. Whiting, R. T. Larsen, and J. M. Shannon. 2020. Determining timing of births and habitat selection to identify lambing period habitat for bighorn sheep. Frontiers in Ecology and Evolution 8:97.

Robinson, R. W., J. C. Whiting, J. M. Shannon, D. D. Olson, J. T. Flinders, T. S. Smith, and R. T. Bowyer. 2019. Habitat use and social mixing between resident and augmented groups of bighorn sheep. Scientific Reports 9:14984.

Robinson, S. E., and F. W. Weckerly. 2010. Grouping patterns and selection of forage by the scimitar-horned oryx (*Oryx dammah*) in the Llano Uplift region of Texas. Southwestern Naturalist 55:510–516.

Rode, K. D., S. D. Farley, and C. T. Robbins. 2006. Sexual dimorphism, reproductive strategy, and human activities determine resource use by brown bears. Ecology 87:2636–2646.

Rodgers, A. R., A. P. Carr, H. L. Beyer, L. Smith, and J. G. Kie. 2007. HRT: home range tools for ArcGIS. Ontario Ministry of Natural Resources, Centre for Northern Forest Ecosystem Research, Thunder Bay, Canada.

Rodgers, P. A., H. Sawyer, T. W. Mong, S. Stephens, and M. J. Kauffman. 2021. Sex-specific migratory behaviors in a temperate ungulate. Ecosphere 12:e03424.

Roffler, G. H., L. G. Adams, and M. Hebblewhite. 2017. Summer habitat selection by Dall's sheep in Wrangell–St. Elias National Park and Preserve, Alaska. Journal of Mammalogy 98:94–105.

Romano, A., S. Salvidio, R. Palozzi, and V. Sbordoni. 2012. Diet of the newt, *Triturus carnifex* (Laurenti, 1786), in the flooded karst sinkhole Pozzo Del Merro, central Italy. Journal of Cave and Karst Studies 74:271–277.

Romey, W. L., and A. C. Wallace. 2007. Sex and the selfish herd: sexual segregation within nonmating whirligig groups. Behavioral Ecology 18:910–915.

Rosas, C. A., D. M. Engle, and J. H. Shaw. 2005. Potential ecological impact of diet selectivity and bison herd composition. Great Plains Research 15:3–13.

Rubenstein, D. I. 1986. Ecology and sociality in horses and zebras. Pages 282–302 in D. I. Rubenstein and R. W. Wrangham, editors. Ecological aspects of social evolution: birds and mammals. Princeton University Press, Princeton, New Jersey, USA.

Rubin, E. S., and V. C. Bleich. 2005. Sexual segregation: a necessary consideration in wildlife conservation. Pages 379–391 in K. E. Ruckstuhl and P. Neuhaus, editors. Sexual segregation in vertebrates: ecology of the two sexes. Cambridge University Press, Cambridge, United Kingdom.

Rubin, E. S., W. M. Boyce, C. J. Stermer, and S. G. Torres. 2002. Bighorn sheep habitat use and selection near an urban environment. Biological Conservation 104:251–263.

Ruckstuhl, K. E. 1998. Foraging behaviour and sexual segregation in bighorn sheep. Animal Behaviour 56:99–106.

Ruckstuhl, K. E. 1999. To synchronise or not to synchronise: a dilemma for young bighorn males? Behaviour 136:805.

Ruckstuhl, K. E. 2007. Sexual segregation in vertebrates: proximate and ultimate causes. Integrative and Comparative Biology 47:245–257.

Ruckstuhl, K. E., and H. Kokko. 2002. Modelling sexual segregation in ungulates: effects of group size, activity budgets and synchrony. Animal Behaviour. 64: 909–914.

Ruckstuhl, K. E., and P. Neuhaus. 2000. Sexual segregation in ungulates: a new approach. Behaviour 137:361–77.

Ruckstuhl, K. E., and P. Neuhaus. 2002. Sexual segregation in ungulates: a comparative test of three hypotheses. Biology Reviews 77:77–96.

Ruckstuhl, K. E., and P. Neuhaus, editors. 2005. Sexual segregation in vertebrates: ecology of the two sexes. Cambridge University Press, Cambridge, United Kingdom.

Ruckstuhl, K. E., and P. Neuhaus. 2009. Activity budgets and sociality in a monomorphic ungulate: the African oryx (*Oryx gazelle*). Canadian Journal of Zoology 87:165–174.

Ruckstuhl, K. E., A. Manica, A. D. C. MacColl, J. G. Pilkington, and T. H. Clutton-Brock. 2006. The effects of castration, sex ratio and population density on social segregation and habitat use in Soay sheep. Behavioral Ecology and Sociobiology 59:694–703.

Ruedas, L. A., J. R. Demboski, and R. V. Sison. 1994. Morphological and ecological variation in *Otopteropus cartilagonodus* Kock 1969 (Mammalia: Chiroptera: Pteropodidae) from Luzon, Philippines. Proceedings of the Biological Society of Washington 107:1–16.

Rughetti, M., and M. Festa-Bianchet. 2011. Seasonal changes in sexual size dimorphism in northern chamois. Journal of Zoology, 284:257–264.

Rutburg, A. T. 1984. Birth synchrony in American (*Bison bison*): response to predation or season? Journal of Mammalogy 65:418–423.

Rutburg, A. T. 1987. Adaptive hypotheses of birth synchrony in ruminants: an interspecific test. American Naturalist 130:692–710.

Sadier, B., J.-J. Delannooy, L. Benedetti, D. L. Bourlès, S. Jailliet, J.-M. Geneste, A.-E. Lebatard, and M. Arnold.

2012. Further constraints on the Chauvet cave artwork elaboration. Proceedings of the National Academy of Sciences of the United States of America 109:8002–8006.

Safi, K. 2008. Social bats: the males' perspective. Journal of Mammalogy 89:1342–1350.

Salton, M., R. Krikwood, D. Slip, and R. Harcourt. 2019. Mechanisms for sex-biased segregation in foraging behaviour by a polygynous marine carnivore. Marine Ecology Progress Series 624:213–226.

Sams, M. G., R. L. Lochmiller, C. W. Qualls, D. M. Leslie Jr., and M. E. Payton. 1996. Physiological correlates of neonatal mortality in an overpopulated herd of white-tailed deer. Journal of Mammalogy 77:179–190.

Santora, T. 2020. Should ecologists treat male and female animals like different species? Scientific American. 2 June 2020. https://www.scientificamerican.com/article /should-ecologists-treat-male-and-female-animals-like -different-species.

Sarmento, W., and J. Berger. 2020. Conservation implications of using an imitation carnivore to assess rarely used refuges as critical habitat features in an alpine ungulate. Peer J:e9296.

Sauer, J. R., and M. S. Boyce. 1983. Density dependence and survival of elk in northwestern Wyoming. Journal of Wildlife Management 47:31–37.

Sawyer, H., J. A. Merkel, A. D. Middleton, S. P. Dwinnell, and K. L. Monteith. 2019. Migratory plasticity is not ubiquitous among large herbivores. Journal of Animal Ecology 88:450–460.

Scarbrough, D. L., and P. R. Krausman. 1988. Sexual segregation in desert mule deer. Southwestern Naturalist 33:157–165.

Schaad, W. S., D. S. Wiseman, R. R. Huber, M. S. Mooring, D. D. Reisig, E. R. Osborne, A. L. Kanalakan, and B. M. Hall. 2005. Sexual segregation in bison: a test of multiple hypotheses. Behaviour 142:897–927.

Schaefer, J. S., and F. Messier. 1995. Habitat selection as a hierarchy: the spatial scales of winter foraging by muskoxen. Ecography18:333–334.

Schaefer, J. S., C. M. Bergman, and S. N. Luttich. 2000. Site fidelity of female caribou at multiple spatial scales. Landscape Ecology 15:731–739.

Schaller, G. B. 1968. Hunting behaviour of the cheetah in the Serengeti National Park, Tanzania. East African Wildlife Journal 6:95–100.

Schaller, G. B. 1972. The Serengeti lion; a study of predator-prey relations. University of Chicago Press, Chicago, Illinois, USA.

Schaller, G. B. 1998. Wildlife of the Tibetan steppe. University of Chicago Press, Chicago, Illinois, USA.

Schaller, G. B., and B. Junrang. 1988. Effects of a snowstorm on Tibetan antelope. Journal of Mammalogy 69:631–634.

Scharnweber, K., M. Plath, and M. Tobler. 2011. Trophic niche segregation between the sexes of two species of livebearing fishes (Poeciliidae). Bulletin of Fish Biology 13:11–20.

Schipper, J., J. S. Chanson, F. Chiozza, N. A. Cox, M. Hoffmann, V. Katariya, J. Lamoreux, A. S. L. Rodrigues, S. N. Stuart, H. J. Temple, et al. 2008. The status of the world's land and marine mammals: diversity, threat, and knowledge. Science 322:225–230.

Schneider, T., and P. M. Kappeler. 2016. Gregarious sexual segregation: the unusual social organization of the Malagasy narrow-striped mongoose (*Mungotictis decemlineata*). Behavioral Ecology and Sociobiology 70:913–926.

Schoener, T. W. 1968. Sizes of feeding territories among birds. Ecology 49:123–141.

Schoener, T. W. 1974. Resource partitioning in ecological communities. Science 185:27–39.

Schroeder, C. A., R. T. Bowyer, V. C. Bleich, and T. R. Stephenson. 2010. Sexual segregation in Sierra Nevada bighorn sheep, *Ovis canadensis sierrae*: ramifications for conservation. Arctic, Antarctic, and Alpine Research 42:476–489.

Schuler, K. L., K. M. Leslie Jr., J. H. Shaw, and E. J. Maichak. 2006. Temporal-spatial distribution of American bison (*Bison bison*) in a tallgrass prairie fire mosaic. Journal of Mammalogy 87:539–544.

Schuppli, C., S. S. Utami Atmoko, K. P. van Schaik, and M. A. van Noordwijk. 2021. The development and complexity in Bornean orangutans. Behavioral Ecology and Sociobiology 75:81.

Schwarm, A., S. Ortmann, H. Hofer, W. J. Streich, E. J. Flach, R. Kuhne, J. Hummel, J. C. Castell, and M. Clauss. 2006. Digestion studies in captive Hippopotamidae: a group of large ungulates with an unusually low metabolic rate. Journal of Animal Physiology and Animal Nutrition 90:300–308.

Schwartz, C. C., K. J. Hundertmark, and T. H. Spraker. 1992. An evaluation of selective bull moose harvest on the Kenai Peninsula, Alaska. Alces 28:1–13.

Schwartz, C. C., W. L. Regelin, and A. W. Franzmann. 1987. Seasonal weight dynamics of moose. Swedish Wildlife Research Supplement 1:301–310.

Scott, J. M., J. A. Wiens, V. Van Horne, and D. D. Goble. 2020. Shepherding nature: the challenge of conservation reliance. Cambridge University Press, Cambridge, United Kingdom.

Selander, R. K. 1966. Sexual dimorphism and differential niche utilization in birds. Condor 68:113–151.

Sells, S. H., M. S. Mitchell, K. M. Podruzny, J. G. Gude, A. C. Keever, D. K. Boyd, T. D. Smucker, A. A. Nelson, T. W. Parks, N. J. Lance, et al. 2021. Evidence of economical territory selection in a cooperative carnivore. Proceedings of the Royal Society of London B 228:20210108.

Senior, P., R. K. Butlin, and J. D. Altringham. 2005. Sex and segregation in temperate bats. Proceedings of the Royal Society of London, B 272:2467–2473.

Sequeira, F., H. Goncalves, M. M. Fari, V. Meneses, and J. W. Arntzen. 2001. Habitat-structure and meteorological parameters influencing the activity and local distribution of the golden-striped salamander, *Chioglossa lusitanica*. Herpetological Journal 11:85–90.

Shallow, J. R. T., M. A. Hurley, K. L. Monteith, and R. T. Bowyer. 2015. Cascading effects of habitat on maternal condition and life-history characteristics of neonatal mule deer. Journal of Mammalogy 96:194–205.

Shank, C. C. 1982. Age-sex differences in the diets of wintering Rocky Mountain bighorn sheep. Ecology 63:627–633.

Shank, C. C. 1985. Inter-and intra-sexual segregation of chamois (*Rupicapra rupicapra*) by altitude and habitat during summer. Zeitschrift für Säugetierkunde 50:117–125.

Shannon, G., B. R. Page, K. J. Duffy, and R. Slotow. 2006a. The consequences of body size dimorphism: are African elephants sexually segregated at the habitat scale? Behaviour 143:1145–1168.

Shannon, G., B. R. Page, K. J. Duffy, and R. Slotow. 2006b. The role of foraging behaviour in the sexual segregation of the African elephant. Oecologia 150:344–354.

Shannon, G., B. R. Page, R. L. Mackey, K. J. Duffy, and R. Slotow. 2008. Activity budgets and sexual segregation in African elephants (*Loxodonta africana*). Journal of Mammalogy 89:467–476.

Shannon, J. M., J. C. Whiting, R. T. Larsen, D. D. Olson, J. T. Flinders, T. S. Smith, and R. T. Bowyer. 2014. Population response of reintroduced bighorn sheep after observed commingling with domestic sheep. European Journal of Wildlife Research 60:737–748.

Shelton, D. E., A. D. Harlin-Cognato, R. L. Honeycutt, and T. M. Markowitz. 2010. Sexual segregation and genetic relatedness in New Zealand. Pages 195–209 *in* B. Würsig and M. Würsig, editors. The dusky dolphin: master acrobat of different shores. Elsevier, Amsterdam, Netherlands.

Shi, J., R. I. M. Dunbar, D. Buckland, and D. Miller. 2003. Daytime activity budgets of feral goats (*Capra hircus*) on the Isle of Rum: influence of season, age, and sex. Canadian Journal of Zoology 81:813–815.

Shi, J., R. I. M. Dunbar, D. Buckland, and D. Miller. 2005. Dynamics of grouping patterns and social segregation in feral goats (*Capra hircus*) on the Isle of Rum, NW Scotland. Mammalia 69:185–199.

Shields, A. V., R. T. Larsen, and J. C. Whiting. 2012. Summer watering patterns of mule deer in the Great Basin Desert, USA: implications of differential use by individuals and the sexes for management of water resources. Scientific World Journal 2012:art.846218.

Shine, R. 1989. Ecological causes for the evolution of sexual dimorphism: a review of the evidence. Quarterly Review of Biology 64:419–460.

Shine, R., and M. Wall. 2005. Ecological divergence between the sexes in reptiles. Pages 221–253 in K. E. Ruckstuhl and P. Neuhaus, editors. Sexual segregation in vertebrates: ecology of the two sexes. Cambridge University Press, Cambridge, United Kingdom.

Shmida, A., S. Lev-Yadun, S. Goubitz, and G. Néeman. 2000. Sexual allocation and gender segregation in *Pinus halepensis*, *P. brutia* and *P. pinea*. Pages 94–104 in G. Néeman and L. Trabauds, editors. Forest ecosystems in the Mediterranean. Backhuys Publishers, Leiden, the Netherlands.

Short, H. L. 1963. Rumen fermentation and energy relationships in the white-tailed deer. Journal of Wildlife Management 28:445–458.

Short, H. L. 1981. Nutrition and metabolism. Pages 99–127 in O. C. Wallmo, editor. Mule and black-tailed deer of North America. University of Nebraska Press, Lincoln, Nebraska, USA.

Short, R. V., and E. Balaban, editors. 1994. The differences between the sexes. Cambridge University Press, Cambridge, United Kingdom.

Sibly, R. M., W. Zuo, A. Kodric-Brown, and J. H. Brown. 2012. Rensch's rule in large herbivorous mammals derived from metabolic scaling. American Naturalist 179:169–177.

Siegfried, W. R. 1979. Vigilance and group size in springbok. Madoqua 12:151–154.

Silvy, N. J., editor. 2020. The wildlife techniques manual. Volumes 1 and 2. Eighth edition. Johns Hopkins University Press, Baltimore, Maryland, USA.

Simberloff, D. 1998. Flagships, umbrellas, and keystones: is single species management passé in the landscape era? Biological Conservation 83:247–257.

Simpson, S. J., N. E. Humphries, and D. W. Sims. 2021. Habitat selection, fine-scale spatial partitioning and sexual segregation in Rajidae, determined using passive acoustic telemetry. Marine Ecology Progress Series 666:115–134.

Sims, D. W. 2005. Differences in habitat selection and reproductive strategies of male and female sharks. Pages 127–147 in K. E. Ruckstuhl and P. Neuhaus, editors. Sexual segregation in vertebrates: ecology of the two sexes. Cambridge University Press, Cambridge, United Kingdom.

Sinclair, A. R. E. 1977. The African buffalo: a study of resource limitation of populations. University of Chicago Press, Chicago, Illinois, USA.

Sinclair, A. R. E. 2000. Adaptation, niche partitioning, and coexistence of African Bovidae: clues to the past. Pages 247–260 in E. S. Vrba and G. B. Schaller, editors. Antelopes, deer, and relatives: fossil record, behavioral ecology, systematics, and conservation. Yale University Press, New Haven, Connecticut, USA.

Singh, N. J., S. Amgalabaatar, and R. P. Reading. 2010a. Temporal dynamics of group size and sexual segregation in ibex. Martin-Luther-Universität Halle Wittenberg, Halle (Saale) 11:315–322.

Singh, N. J., C. Nomefant, N. G. Yoccoz, and S. D. Côté. 2010b. Sexual segregation in Eurasian wild sheep. Behavioral Ecology 21:410–418.

Siuta, A. 2006. Functional stomach chamber strategy of red deer in relation to sex and mating season. Medycyna Weterynaryjna 62:773–777.

Skogland, T. 1989. Comparative social organization of wild reindeer in relation to food, mates and predator avoidance. Advances in Ethology 29:1–74.

Slatkin, M. 1984. Ecological causes of sexual dimorphism. Evolution 38:622–630.

Smallwood, J. A. 1987. Sexual segregation by habitat in American kestrels wintering in southcentral Florida: vegetative structure and responses to differential prey availability. Condor 89:842–849.

Smiley, R. A., C. D. Rittenhouse, T. W. Mong, and K. L. Monteith. 2020. Assessing nutritional condition of mule deer using a photographic index. Wildlife Society Bulletin 44:208–213.

Smit, I. P. J., C. C. Grant, and J. Whyte. 2007. Landscape-scale sexual segregation in the dry season distribution and resource utilization of elephants in Kruger National Park, South Africa. Diversity and Distributions 13:225–236.

Smith, D. W., and R. O. Peterson. 2021. Intended and unintended consequences of wolf restoration to Yellowstone and Isle Royale National Parks. Conservation Science and Practice 2021:e413.

Smith, D. W., D. R. Stahler, and D. R. MacNulty, editors. 2020. Yellowstone wolves: science and discovery in the world's first nation park. University of Chicago Press, Chicago, Illinois, USA.

Smith, K. T., J. L. Beck, and C. P. Kirol. 2018. Reproductive state leads to intraspecific habitat partitioning and survival differences in greater sage-grouse: implication for conservation. Wildlife Research 45:119–131.

Smith, W. P. 1987. Maternal defense in Columbian white-tailed deer: when is it worth it? American Naturalist 130:310–316.

Smythe, N. 1970. On the existence of "pursuit invitation" signals in mammals. American Naturalist 104:491–494.

Smythe, N. 1977. The function of mammalian alarm advertising: social signals or pursuit invitation? American Naturalist 111:191–194.

Soulsbury, C. D., M. Kervinen, and C. Lebigre. 2014. Sexual size dimorphism and the strength of sexual selection in mammals and birds. Evolutionary Ecology Research 16:63–67.

Southwick, R., and T. Allen. 2010. Expenditures, economic impacts and conservation contributions of hunters in the United States. Pages 308–313 in World symposium: ecologic and economic benefits of hunting. World Forum on the Future of Sport Shooting Activities, Windhoek, Namibia.

Spaeth, D. F., R. T. Bowyer, T. R. Stephenson, and P. S. Barboza. 2004. Sexual segregation in moose Alces alces: an experimental manipulation of foraging behaviour. Wildlife Biology 10:59–72.

Spaeth, D. F., R. T. Bowyer, T. R. Stephenson, P. S. Barboza, and V. Van Ballenberghe. 2002. Nutritional quality of willows for moose: effects of twig age and diameter. Alces 38:143–154.

Spaeth, D. F., K. J. Hundertmark, R. T. Bowyer, P. S. Barboza, T. R. Stephenson, and R. O. Peterson. 2001. Incisor arcades of Alaskan moose: Is dimorphism related to sexual segregation? Alces 37:217–226.

Sparks, D. P., and J. C. Malechek. 1968. Estimating percentage dry weight in diets using a microscopic technique. Journal of Range Management 21:264–265.

Spidal, A. B., and M. D. Johnson. 2016. Sexual habitat segregation in migrant warblers along a shade gradient of Jamaican coffee farms. Journal of Caribbean Ornithology 29:37–42.

Sprogis, K. R., F. Christiansen, H. C. Raudino, H. T. Kobryn, R. S. Wells, and L. Bejdr. 2018. Sex-specific differences in the seasonal habitat use of a coastal dolphin population. Biodiversity and Conservation 27:3637–3656.

Sprogis, K. R., K. H. Pollock, H. C. Ravdino, S. J. Alley, A. M. Kopps, O. Manlik, J. A. Tyne, and L. Bejder. 2016. Sex-specific patterns in abundance, temporary emigration and survival of Indo-Pacific bottlenose dolphins (Tursiops aduncus) in coastal and estuarine waters. Frontiers in Marine Science 3:art.12.

Staines, B. W. 1977. Factors affecting the seasonal distribution of red deer (Cervus elaphus, L.) in Glen Dye, North-East Scotland. Annals of Applied Biology 87:495–512.

Staines, B. W., J. M. Crisp, and T. Parish. 1982. Differences in the quality of food eaten by red deer stags and hinds in winter. Journal of Applied Ecology 19:65–77.

Staniland, I. J. 2005. Sexual segregation in seals. Pages 53–73 in K. E. Ruckstuhl and P. Neuhaus, editors. Sexual segregation in vertebrates: ecology of the two sexes. Cambridge University Press, Cambridge, United Kingdom.

Staniland, I. J., and S. L. Robinson. 2008. Segregation between the sexes: Antarctic fur seals, Arctocephalus gazelle, foraging at South Georgia. Animal Behaviour 75:1581–1590.

Stankowitch, T., and T. Caro. 2009. Evolution of weaponry in female bovids. Proceedings of the Royale Society of London B 276:4329–4334.

Stankowitch, T., and R. G. Coss. 2008. Alarm walking in Columbian black-tailed deer: its characterization and possible antipredatory signaling functions. Journal of Mammalogy 89:636–645.

Stark, L. R., D. N. McLetchie, and B. D. Mishler. 2005. Sex expression, plant size, and spatial segregation in the desert moss Syntrichia caninervis. Bryologist 108:183–193.

Stauss, C., S. Bearhop, T. W. Bodey, S. Garthe, C. Gunn, W. J. Grecian, R. Inger, M. E. Night, J. Newton, S. C. Patrick, et al. 2012. Sex-specific foraging behaviour in northern gannets Morus bassanus: incidence and implications. Marine Ecology Progress Series 457:151–162.

Stearns, S. C. 1977. The evolution of life history traits. A critique of the theory and a review of the data. Annual Review of Ecology and Systematics 8:145–171.

Steiniger, S., and A. J. S. Hunter. 2012. OpenJUMP HoRAE: a free GIS and toolbox for home-range analysis. Wildlife Society Bulletin 36:600–608.

Stephenson, T. R., V. Van Ballenberghe, J. M. Peek, and J. G. MacCracken. 2006. Spatiotemporal constraints on moose habitat and carrying capacity in coastal Alaska: vegetation succession and climate. Range Ecology and Management 59:359–372.

Stevens, C. E. 1988. Comparative physiology of the vertebrate digestive system. Cambridge University Press, New York, New York, USA.

Stewart, K. M., R. T. Bowyer, B. L. Dick, B. K. Johnson, and J. G. Kie. 2005. Density-dependent effects on physical condition and reproduction in North American elk: an experimental test. Oecologia 143:85–93.

Stewart, K. M., R. T. Bowyer, J. G. Kie, N. J. Cimon, and B. K. Johnson. 2002. Temporospatial distributions of elk, mule deer, and cattle: resource partitioning and competitive displacement. Journal of Mammalogy 83:229–244.

Stewart, K. M., R. T. Bowyer, J. G. Kie, B. L. Dick, and M. Ben-David. 2003a. Niche partitioning among mule deer, elk, and cattle: do stable isotopes reflect dietary niche? Écoscience 10:297–302.

Stewart, K. M., R. T. Bowyer, J. G. Kie, B. L. Dick, and R. W. Ruess. 2009. Population density of North American elk: effects on plant diversity. Oecologia 161:303–312.

Stewart, K. M., R. T. Bowyer, J. G. Kie, and W. C. Gasaway. 2000. Antler size relative to body mass in moose: tradeoffs associated with reproduction. Alces 36:77–83.

Stewart, K. M., R. T. Bowyer, J. G. Kie, and M. A. Hurley. 2010. Spatial distributions of mule deer and North American elk: resource partitioning in a sage-steppe environment. American Midland Naturalist 163:400–412.

Stewart, K. M., R. T. Bowyer, R. W. Ruess, B. L. Dick, and J. G. Kie. 2006. Herbivore optimization by North American elk: consequences for theory and manage-ment. Wildlife Monographs 167:1–24.

Stewart, K. M., R. T. Bowyer, and P. J. Weisberg. 2011. Spatial use of landscapes. Pages 181–217 in D. G. Hewitt, editor. Biology and management of white-tailed deer. CRC Press, Boca Raton, Florida, USA.

Stewart, K. M., T. E. Fulbright, D. L. Drawe, and R. T. Bowyer. 2003b. Sexual segregation in white-tailed deer: responses to habitat manipulations. Wildlife Society Bulletin 31:1210–1217.

Stewart, K. M., D. R. Walsh, J. G. Kie, B. L. Dick, and R. T. Bowyer. 2015. Sexual segregation in North American elk: the role of density dependence. Ecology and Evolution 5:709–721.

Steyaert, S. J. M. G., J. Kindberg, J. W. Swenson, and A. Zedrosser. 2013. Male reproductive strategy explains spatiotemporal segregation in brown bears. Journal of Animal Ecology 82:836–845.

Stokke, S. 1999. Sex differences in feeding-patch choice in a megaherbivore: elephants in Chobe National Park, Botswana. Canadian Journal of Zoology 77:1723–1732.

Stokke, S., and J. T. du Toit. 2000. Sex and size related differences in the dry season feeding patterns of elephants in Chobe National Park, Botswana. Ecography 23:70–80.

Stokke, S., and J. T. du Toit. 2002. Sexual segregation in habitat use by elephants in Chobe National Park, Botswana. African Journal of Ecology 40:360–371.

Stokke, S., and J. T. du Toit. 2014. The Chobe elephants: one species, two niches. Pages 104–117 in Elephants and savanna woodland ecosystems: a study from Chobe National Park, Botswana. John Wiley and Sons, Hoboken New Jersey, USA.

Stone, D. B., J. A. Martin, B. S. Cohen, T. J. Prbyl, C. Killmaster, and K. B. Miller. 2019. Intraspecific temporal resource partitioning at white-tailed deer feeding sites. Current Zoology 65:139–146.

Storer, R. W. 1966. Sexual dimorphism and food habits in three North American accipiters. Auk 83:423–436.

Street, G. M., F. W. Weckerly, and S. Schwinning. 2013. Modeling forage mediated aggregation in a gregarious ruminant. Oikos 122:922–929.

Stubbs, M. 1977. Density dependence in life-cycles of animals and its importance in K- and r-selected strategies. Journal of Animal Ecology 56:677–688.

Sukumar, R. 1989. The Asian elephant: ecology and management. Cambridge University Press, Cambridge, United Kingdom.

Sukumar, R., and M. Gadgil. 1988. Male-female differences in foraging on crops by Asian elephants. Animal Behaviour 36:1233–1235.

Suominen, O., K. Dannell, and R. Bergström. 1999a. Moose, trees, and ground-living invertebrates: indirect interactions in Swedish pine forests. Oikos 84:215–226.

Suominen, O., K. Dannell, and J. P. Bryant. 1999b. Indirect effects of mammalian browsers on vegetation and ground-dwelling insects in an Alaskan flood plain. Écoscience 6:505–510.

Suraci, J. P., M. Clinchy, L. M. Dill, D. Robers, and L. Y. Zanette. 2016. Fear of large carnivores causes a trophic cascade. Nature Communications 7:1–7.

Székely, T., J. N. Webb, and I. C. Cuthill. 2000. Mating patterns, sexual selection and parental care: an integrative approach. Pages 159–185 in M. Apollonio, M. Festa-Bianchet, and D. Mainardi, editors. Vertebrate mating systems. World Scientific, New Jersey, USA.

Szpak, P., M.-H. Julien, T. C. A. Royle, J. M. Savelle, D. Y. Yang, and M. P. Richards. 2019. Sexual differences in the foraging ecology of the 19th century beluga whales (*Delphinapterus leucas*) from the Canadian High Arctic. Marine Mammal Science 36:451–471.

Takada, H., and M. Minami. 2021. Open habitats promote female group formation in a solitary ungulate: the Japanese serow. Behavioral Ecology and Sociobiology 75:60.

Tarango, L. A., P. R. Krausman, and R. Valdex. 2002. Habitat use by desert bighorn sheep in Sonora, Mexico. Pirineos 157:219–226.

Telfer, E. S. 1967. Comparison of a deer yard and a moose yard in Nova Scotia. Canadian Journal of Zoology 45:485–490.

Telfer, E. S. 1978. Cervid distribution, browse and snow cover in Alberta. Journal of Wildlife Management 42:352–361.

Terborgh, J., R. D. Holt, and J. A. Estes. 2010. Trophic cascades: what they are, how they work, and why they matter. Pages 1–18 in J. Terborgh and J. A. Estes, editors. Trophic cascades: predators, prey, and the changing dynamics of Nature. Island Press, Washington, DC, USA.

Terborgh, J., L. L. Lopez, P. Nuñez, M. Rao, G. Shahabud-din, G. Orihuela, M. Riveros, R. Ascanio, G. H. Alder,

T. D. Lambert, and L. Balbas. 2001. Ecological meltdown in predator-free forest fragments. Science 294:1923–1926.

Tettamanti, F., and V. A. Viblanc. 2014. Influences on mating group composition on the behavioural time-budget of male and female alpine ibex (*Capra ibex*) during the rut. PLoS ONE 9:e86004.

Thalmann, J. C., R. T. Bowyer, K. A. Aho, F. W. Weckerly, and D. R. McCullough. 2015. Antler and body size in black-tailed deer: an analysis of cohort effects. Advances in Ecology 2015:art.156041.

Thirgood, S. J. 1996. Ecological factors influencing sexual segregation and group size in fallow deer (*Dama dama*). Journal of Zoology 239:783–797.

Thomas, D. L., and E. J. Taylor. 2006. Study designs and tests for comparing resource use and availability II. Journal of Wildlife Management, 70:324–336.

Thomas, R. C., and C. B. Schultz. 2016. Resource selection in an endangered butterfly: females select native nectar species. Journal of Wildlife Management 80:171–180.

Thompson, D. P., and P. S. Barboza. 2017. Seasonal energy and protein requirements for Siberian reindeer (*Rangifer tarandus*). Journal of Mammalogy 98:1558–1567.

Tian, J., J. Du, Z. Lu, Y. Li, D. Li, J. Han, Z. Wang, and X. Guan. 2021. Differences in the fecal microbiota due to the sexual niche segregation of captive gentoo penguins *Pygoscelis papua*. Polar Biology 44:473–482.

Tilley, J. M. A., and R. A. Terry. 1963. A two-stage technique for the in vitro digestion of forage crops. Journal of the British Grassland Society 18:104–111.

Toïgo, C., and J.-M. Gaillard. 2003. Causes of sex-biased adult survival in ungulates: sexual dimorphism, mating tactic or environment harshness? Oikos 101:378–384.

Tollefson, T. N., L. A. Shipley, W. L. Myers, N. Dasgupta. 2011. Forage quality's influence on mule deer fawns. Journal of Wildlife Management 75:919–928.

Tollefson, T. N., L. A. Shipley, W. L. Myers, D. H. Keisler, and N. Dasgupta. 2010. Influence of summer and autumn nutrition on body condition and reproduction in lactating mule deer. Journal of Wildlife Management 74:974–986.

Torices, R., A. Afonso, A. A. Aderberg, J. M. Gomez, and M. Méndez. 2019. Architectural traits constrain the evolution of unisexual flowers and sexual segregation within inflorescences: an interspecific approach. Peer Community in Evolutionary Biology. https://doi.org/10.1101/356147.

Toupin, B., J. Huot, and M. Manseau. 1996. Effect of insect harassment on the behaviour of the Rivière George caribou herd. Arctic 49:375–382.

Trefethen, J. B. 1975. An American crusade for wildlife. Boone and Crockett Club, Alexandria, Virginia, USA.

Trivers, R. L. 1972. Parental investment and sexual selection, Pages 136–179 in B. Campbell, editor. Sexual selection and the decent of man. Aldine, Chicago, Illinois, USA.

Tucker, S., W. D. Bowen, and S. J. Iverson. 2007. Dimensions of diet segregation in grey seals *Halichoerus grypus* revealed through stable isotopes of carbon (δ13C) and nitrogen (δ15N). Marine Ecology Progress Series 339:271–282.

Tuomainen, A., T. Valtonen, and D. Benesh. 2015. Sexual segregation of *Echinorhynchus borealis* von Linstow, 1901 (Acanthocephala) in the gut of burbot (*Lota lota* Linnaeus). Folia Parasitologica 62:061.

Turner, M. M., A. P. Rockhill, C. S. Deperno, J. A. Jenks, R. W. Klaver, A. R. Jarding, R. W. Grovenburg, and K. H. Pollock. 2011. Evaluating the effect of predators on white-tailed deer: movement and diets of coyotes. Journal of Wildlife Management 75:905–912.

Turner, W. C., A. E. Jolles, and N. Owen-Smith. 2005. Alternating sexual segregation during the mating season by male African buffalo (*Syncerus caffer*). Journal of Zoology, 267:291–299.

Uccheddu, S., G. Body, R. B. Weladji, Ø. Holand, and M. Nieminen. 2015. Foraging competition in larger groups overrides harassment avoidance benefits in female reindeer (*Rangifer tarandus*). Oecologia 179:711–718.

Unterthiner, S., F. Ferretti, L. Rossi, and S. Lovari. 2012. Sexual and seasonal differences of space use in alpine chamois. Ethology Ecology and Evolution 24:257–274.

Urban, M. C. 2015. Accelerating extinction risk from climate change. Science 348:571–573.

Valdex, J., K. Klop-Toker, M. P. Stockwell, S. Clulow, J. Clulow, and M. J. Mahony. 2016. Microhabitat selection varies by sex and age class in the endangered green and golden bell frog *Litoria aurea*. Australian Zoologist 38:223–234.

Valentini, A., F. Pompanon, and P. Taberlet. 2009. DNA barcoding for ecologists. Trends in Ecology and Evolution 24:110–117.

Vampé, C., J.-M. Gaillard, P. Kjellander, A. Mysterud, P. Magnien, D. Delorme, G. Van Laere, F. Klein, O. Liberg, and A. J. M. Hewison. 2007. Antler size provides an honest signal of male phenotypic quality in roe deer. American Naturalist 169:481–493.

Vampé, C., P. Kjellander, M. Galan, J.-F. Cosson, S. Aulagnier, O. Liberg, and A. J. M. Hewison. 2008. Mating system, sexual dimorphism, and the opportunity for sexual selection in a territorial ungulate. Behavioral Ecology 19:309–316.

Van Ballenberghe, V., and W. B. Ballard. 1994. Limitation and regulation of moose populations: the role of predation. Canadian Journal of Zoology 72:2071–2077.

Vander Wal, E., P. C. Paquet, F. Messier, and P. D. McLoughlin. 2013. Effects of phenology and sex on social proximity in a gregarious ungulate. Canadian Journal of Zoology 91:601–609.

van Dijk, J., L. Gustavsen, A. Mysterud, R. May, O. Flagstad, H. Brøseth, R. Andersen, R. Andersen, H. Steen, and A. Landa. 2008. Diet shift of a facultative scavenger, the wolverine, following recolonization of wolves. Journal of Animal Ecology 77:1183–1190.

Van Drunen, W. E., and M. E. Dorken. 2014. Wind pollination, clonality, and the evolutionary maintenance of spatial segregation of the sexes. Evolutionary Ecology 28:1121–1138.

Van Horne, B. 1983. Density as a misleading indicator of habitat quality. Journal of Wildlife Management 47:893–901.

Van Soest, P. J. 1994. Nutritional Ecology of the Ruminant. Second edition. Cornell University Press, Ithaca, New York, USA.

van Toor, M. L., C. Jaberg, and K. Safi. 2011. Integrating sex-specific habitat use for conservation using habitat suitability models. Animal Conservation 14:512–520.

Van Vliet, N., J. Fa, and R. Nasi. 2015. Managing hunting under uncertainty: from one-off ecological indicators to resilience approaches in assessing the sustainability of bushmeat hunting. Ecology and Society 20:7.

van Wieren, S. E., and S. de Bie. 1979. Sexual segregation in wild reindeer on Edgeoya (Svalbard). Proceedings of the International Reindeer and Caribou Symposium 2:550–553.

Vásquez-Castillo, S., I. A. Hinojosa, N. Colin, A. A. Poblete, and K. Górski. 2021. The presence of kelp *Lessonia trabeculata* drives isotopic niche segregation of redspotted catshark *Schroederichthys chilensis*. Estuarine, Coastal, and Shelf Science 258:107435.

Verhulst, P. F. 1838. Notice sur la loi que la populations suit dans son accroissement. Correspondence Mathematiqe et Physique 10:113–121. Original not seen; citation from Hutchinson (1978).

Verme, L. J. 1988. Niche selection by male white-tailed deer: an alternative hypothesis. Wildlife Society Bulletin 16:448–451.

Viana, D. S., J. E. Granados, P. Fandos, J. M. Pérez, F. J. Cano-Manuel, D. Burón, G. Fandos, M. Á. P. Aguado, J. Figuerola, and R. C. Soriguer. 2018. Linking seasonal home range size with habitat selection and movement in a mountain ungulate. Movement Ecology 6:1.

Villaret, J. C., and R. Bon. 1995. Social and spatial segregation in alpine ibex (*Capra ibex*) in Bargy, French Alps. Ethology 101:291–300.

Villaret, J. C., R. Bon, and A. Rivet. 1997. Sexual segregation of habitat by the alpine ibex in the French Alps. Journal of Mammalogy 78:1273–1281.

Villepique, J. T., B. M. Pierce, V. C. Bleich, A. Andic, and R. T. Bowyer. 2015. Resource selection by an endangered ungulate: a test of predator-induced range abandonment. Advances in Ecology 2015:art.357080.

Villerette, N., R. Helder, J.-M. Angibault, B. Cargnelutti, J.-F. Gerard. 2006a. Sexual segregation in fallow deer: are mixed-sex groups especially unstable because of asynchrony between the sexes? Comptes Rendus Biologies 329:551–558.

Villerette, N., C. Marchal, O. Pays, D. Delome, and J.-F. Gerard. 2006b. Do the sexes tend to segregate in roe deer in agricultural environments? An analysis of group composition. Canadian Journal of Zoology 84:787–796.

Vitousek, P. M., H. A. Mooney, J. Lubencho, and J. M. Melillo. 1997. Human domination of Earth's ecosystems. Science 277:292–499.

Volterra, V. 1926. Fluctuations in the abundance of a species considered mathematically. Nature 118:558–560.

Walker, B. D., K. L. Parker, and M. P. Gillingham. 2006. Behaviour, habitat associations, and intrasexual differences in female Stone's sheep. Canadian Journal of Zoology 84:1187–1201.

Walter, W. D. 2014. Use of stable isotopes to identify dietary difference across subpopulations and sex for a free-ranging generalist herbivore. Isotopes in Environmental and Health studies 50:399–413.

Walters, C. J., and R. Hilborn. 1978. Ecological optimization and adaptive management. Annual Review of Ecology and Systematics 9:157–188.

Walther, F. R. 1984. Communication and expression in hooved mammals. Indiana University Press, Bloomington, Indiana, USA.

Wang, M., J. Alves, A. A. da Silva, W. Yang, and K. E. Ruckstuhl. 2008. The effect of male age on patterns of sexual segregation in Siberian ibex. Scientific Reports 8:art.13095.

Wang, M., D. Blank, Y. Wang, W. Xu, W. Yang, and J. Alves. 2019. Seasonal changes in the sexual segregation patterns of Marco Polo sheep in Taxkorgan Nature Reserve. Journal of Ethology 37:203–211.

Watson, A., and B. W. Staines. 1978. Differences in the quality of wintering areas used by male and female red deer (*Cervus elaphus*) in Aberdeenshire. Journal of Zoology, 286:544–550.

Watts, D. P. 2005. Sexual segregation in non-human primates. Pages 327–347 *in* K. E. Ruckstuhl and P. Neuhaus, editors. Sexual segregation in vertebrates: ecology of the two sexes. Cambridge University Press, Cambridge, United Kingdom.

Wearmouth, V. J., and D. W. Sims. 2008. Sexual segregation in marine fish, reptiles, birds and mammals: behavior patterns, mechanisms and conservation implications. Advances in Marine Biology 54:107–168.

Wearmouth, V. J., E. J. Southall, D. Morritt, R. C. Thompson, I. C. Cuthill, J. C. Partridge, and D. W. Sims. 2012. Year-round sexual harassment as a behavioral mediator of vertebrate population dynamics. Ecological Monographs 82:35–356.

Webster, T., S. Dawson, and E. Slooten. 2009. Evidence of sex segregation in Hector's dolphin (*Cephalorhynchus*). Aquatic Mammals 35:212–219.

Weckerly, B. [F. W.]. 2017. Population ecology of Roosevelt elk: conservation and management in Redwood National and State Parks. University of Nevada Press, Reno, Nevada, USA.

Weckerly, F. W. 1993. Intersexual resource partitioning in black-tailed deer: a test of the body size hypothesis. Journal of Wildlife Management 57:475–494.

Weckerly, F. W. 1998a. Sexual segregation and competition in Roosevelt elk. Northwestern Naturalist 79:113–118.

Weckerly, F. W. 1998b. Sexual-size dimorphism: influence of mass and mating systems in the most dimorphic mammals. Journal of Mammalogy 79:33–52.

Weckerly, F. W. 2001. Are large male Roosevelt elk less social because of aggression? Journal of Mammalogy 82:414–421.

Weckerly, F. W. 2010. Allometric scaling of rumen-reticulum capacity in white-tailed deer. Journal of Zoology 280:41–48.

Weckerly, F. W. 2020. Frequency and density associated grouping patterns of male Roosevelt elk. Frontiers in Ecology and Evolution 8:204.

Weckerly, F. W., and J. P. Nelson. 1990. Age and sex differences of white-tailed deer diet composition, quality, and calcium. Journal of Wildlife Management 59:816–823.

Weckerly, F. [W.], K. McFarland, M. Ricca, and K. Meyer. 2004. Roosevelt elk density and social segregation: foraging behavior and females avoiding larger groups of males. American Midland Naturalist 152:386–399.

Weckerly, F. W., M. A. Ricca, and K. P. Meyer. 2001. Sexual segregation in Roosevelt elk: cropping rates and aggression in mixed-sex groups. Journal of Mammalogy 82:825–835.

Weimerskirch, H., M. Le Corre, Y. Ropert-Coudert, A. Kato, and F. Marsac. 2006. Sex-specific foraging behaviour in a seabird with reversed sexual dimorphism: the red-footed booby. Oecologia 146:681–691.

Weixelman, D. A., R. T. Bowyer, and V. Van Ballenberghe. 1998. Diet selection by Alaskan moose during winter: effects of fire and forest succession. Alces 34:213–238.

Weladji, R. B., G. Body, Ø. Holand, X. Meng, and M. Nieminen. 2017. Temporal variation in the operational sex ratio and male mating behaviours in reindeer (*Rangifer tarandus*). Behavioural Processes 140:96–103.

Westgate, M. J., G. E. Likens, and D. B. Lindenmayer. 2013. Adaptive management of biological systems: a review. Biological Conservation 158:128–139.

White, D. R., L. Betzig, M. Borgerhoff Mulder, G. Chick, J. Hartung, W. Irons, B. S. Low, and K. F. Otterbein. 1988. Rethinking polygyny: co-wives, codes, and cultural systems. Current Anthropology 29:579–572.

White, K. R., G. M. Koehler, D. T. Maletzke, and R. B. Wielgus. 2011. Differential prey use by male and female cougars in Washington. Journal of Wildlife Management 75:1115–1120.

White, K. S. 2006. Seasonal and sex-specific variation in terrain use and movement patterns of mountain goats in southeastern Alaska. Biennial Symposium of Northern Wild Sheep and Goat Council 15:183–193.

White, K. S., J. W. Testa, and J. Berger. 2001. Behavioral and ecologic effects of differential predation pressure on moose in Alaska. Journal of Mammalogy 82:422–429.

White, R. G. 1983. Foraging patterns and their multiplier effect on productivity of northern ungulates. Oikos 40:377–384.

White, R. G., and J. R. Luick. 1984. Plasticity and constraints in the lactational strategy of reindeer and caribou. Journal of Zoology 51:215–232.

Whiteside, M. A., J. O. van Horik, E. J. Langley, C. E. Beadsworth, L. A. Capstick, and J. R. Madden. 2018. Size dimorphism and sexual segregation in pheasants: tests of three competing hypotheses. PeerJ 6:e5674.

Whiteside, M. A., J. O. van Horik, E. J. Langley, C. E. Beadsworth, L. A. Capstick, and J. R. Madden. 2019. Patterns of association at feeder stations for common pheasants released into the wild: sexual segregation by space and time. Ibis 161:325–336.

Whiteside, M. A., J. O. van Horik, E. J. Langley, C. E. Beadsworth, P. R. Laker, and J. R. Madden. 2017. Differences in social preference between the sexes during ontogeny drive segregation in a precocial species. Behavioral Ecology and Sociobiology 71:103.

Whiting, J. C., V. C. Bleich, R. T. Bowyer, and R. T. Larsen. 2012. Water availability and bighorn sheep: life-history characteristics and persistence of populations. Pages 127–158 in J. A. Daniels, editor. Advances in environmental research. Volume 21. Nova Science Publishers, Inc., Hauppauge, New York, USA.

Whiting, J. C., R. T. Bowyer, J. T. Flinders, V. C. Bleich, and J. G. Kie. 2010. Sexual segregation and use of water by bighorn sheep: implications for conservation. Animal Conservation 13:541–548.

Wiedmann, B. P., and V. C. Bleich. 2014. Demographic responses of bighorn sheep to reactional activities: a trial of a trail. Wildlife Society Bulletin 38:773–782.

Wielgus, R. B. 2017. Resource competition and apparent competition in declining mule deer (*Odocoileus hemionus*). Canadian Journal of Zoology 95:499–504.

Wielgus, R. B., and F. L. Bunnell. 1995. Tests of hypotheses for sexual segregation in grizzly bears. Journal of Wildlife Management 59:552–560.

Williams, C. T., M. Klaassen, B. M. Barnes, C. L. Buck, W. Arnold, S. Giroud, S. G. Vetter, and R. Ruff. 2017. Seasonal reproductive tactics: annual timing and the capital-to-income breeder continuum. Philosophical Transactions of the Royal Society B 372:20160250.

Williams, G. C. 1966. Adaptation and natural selection: a critique of some current evolutionary thought. Princeton University Press, Princeton, New Jersey, USA.

Willisch, C. S., I. Biebach, U. Koller, T. Bucher, N. Marreros, M.-P. Ryser-Degiorgis, L. F. Keller, and P. Neuhaus. 2012. Male reproductive pattern in a polygynous ungulate with a slow life-history: the role of age, social status, and alternative mating tactics. Evolutionary Ecology 26:187–206.

Wilson, A. J., M. B. Morrissey, M. J. Adams, C. A. Walling, F. E. Guinness, J. M. Pemberton, T. H. Clutton-Brock, and L. E. B. Kruuk. 2011. Indirect genetics effects and evolutionary constraint: an analysis of social dominance in red deer, *Cervus elaphus*. Journal of Evolutionary Biology 24:772–783.

Winkler, D. E., and T. M. Kaiser. 2011. A case study of seasonal, sexual and ontogenetic divergence in the feeding behaviour of the moose (*Alces alces* Linné, 1758). Verhandlungen des Naturwissenschaftlichen Vereins in Hamburg 46:331–348.

Wirtz, P., and P. Kaiser. 1988. Sex differences and seasonal variation in habitat choice in a high density population of waterbuck, *Kobus ellipsiprymnus* (Bovidae). Zeitschrift für Säugetierkunde 53:162–169.

Wolf, J. B. W., G. Kauermann, and F. Trillmich. 2005. Males in the shade: habitat use and sexual segregation in the Galápagos sea lion (*Zalophus californianus wollebaeki*). Behavioral Ecology and Sociobiology 59:293–302.

Woodland, D. J., Z. Jaafar, and M. Knight. 1980. The "pursuit deterrent" function of alarm signals. American Naturalist 115:748–753.

Xavier, J. C., P. N. Trathan, F. R. Ceia, G. A. Traling, S. Adlard, D. Fox, E. W. J. Edwards, R. P. Viera, R. Medeiros, et al. 2017. Sexual and individual foraging segregation in gentoo penguins *Pygoscelis papua* from the Southern Ocean during an abnormal winter. PLoS ONE 12:e0174850.

Xu, F., M. Ma, W. Yang, D. Blank, and Y. Wu. 2012. Test of the activity budget hypothesis on Asiatic ibex in Tian Shan Mountains of Xinjiang, China. European Journal of Wildlife Research 58:71–75.

Yan, W.-B., Z.-G. Zeng, H.-S. Gong, X.-B. He, X-Y. Liu, K.-C. Si, and Y.-L. Song. 2016. Habitat use and selection by takin in the Qinling Mountains, China. Wildlife Research 43:671–680.

Yan, W.-B., X.-G. Zing, D. Pan, T.-J. Wang, Q. Zang, Y.-N. Fu, X.-M. Lin, and Y.-L. Song. 2013. Scale-depending habitat selection by reintroduced Eld's deer (*Cervus eldi*) in a human-dominated landscape. Wildlife Research 40:217–227.

Yearsley, J. M., and F. J. Pérez-Barbería 2005. Does the activity budget hypothesis explain sexual segregation in ungulates? Animal Behaviour 69:257–267.

Young, T. P., and L. A. Isbell. 1991. Sex differences in giraffe feeding ecology: energetic and social constraints. Ethology 87:79–89.

Žák, J., M. Prchalová, M. Šmejkal, P. Blabolil, M. Vašek, J. Maténa, M. Řiha, J. Peterka, J. Sed'q, and J. Kubečka. 2020. Sexual segregation in European cyprinids: consequences of response to predation risk influenced by sexual size dimorphism. Hydrobiologia 847:1439–1451.

Zalewski, A. 2007. Does size dimorphism reduce competition between sexes? The diet of male and female pine martens at local and wider geographical scales. Acta Theriologica 52: 237–250.

Zhao, J. J., Z. L. Wang, X. M. Chen, and Y. Chen. 2013. Sex differences in piercing-sucking sites on leaves of *Ligustrum lucidum* (Oleaceae) infested by the Chinese white wax scale insect, *Ericerus pela* (Chavannes) (Hemiptera: Coccidae). Neotropical Entomology 42:158–163.

Zhu, Z., Y. Sun, F. Zhu, Z. Liu, R. Pan, L. Teng, and S. Guo. 2020. Seasonal variation and sexual dimorphism in the microbiota of wild blue sheep (*Pseudois nayaur*). Frontiers in Microbiology 11:1260.

Zimmerman, G. M., H. Goetz, and W. P. Mielke. 1985. Use of an improved statistical method for group comparisons to study effects of prairie fire. Ecology 66:606–611.

Zimmerman, T. J., J. A. Jenks, and D. M. Leslie Jr. 2006. Gastrointestinal morphology of female white-tailed and mule deer: effects of fire, reproduction, and feeding type. Journal of Mammalogy 87:598–605.

Zuk, M., and K. A. McKean. 1996. Sex differences in parasite infections: patterns and processes. International Journal for Parasitology 26:1009–1024.

Zweifel-Schielly, B., Y. Leuenberger, M. Kreuzer, and W. Suter. 2011. A herbivore's food landscape: seasonal dynamics and nutritional implication of diet selection by a red deer population in contrasting alpine habitats. Journal of Zoology 286:68–80.

Index

activity-budget hypothesis, 86–88, 92

activity patterns, in sexual segregation, 86–90, 91; diel cycle, 86, 87, 92; in social segregation, 22, 62, 87; in spatial segregation, 50, 52, 72, 88, 90; tests for, 100

activity-patterns hypothesis, 86–91, 128, case study, 88–90, 92; criticisms, 90, 91–92

adaptation: antipredator behaviors, 36, 39; definition, 21; phylogenetic inertia, 21; in ruminal microbes, 30; sexual segregation as, 21, 23, 52, 59, 75, 79, 82; sociality as, 25

Africa: deforestation in, 11; ungulate distribution in, 2

African antelopes (Bovidae), 2, 25, 29; giant eland (*Taurotragus derbianus*), 2; royal (*Neotragus pygmaeus*), 2, 25, 29

African buffalo (*Syncerus caffer*), 18, 39, 45, 49, 84

age factors: in aggression, 85; in male harvesting, 106; in predation and nutritional risks relationship, 102; in reproduction, 113; in sexual segregation, 74

aggression: activity patterns and, 91; age factors, 85; antipredator, 39; group size and, 34, 39, 85, 86; intersexual, 85–86, 128; male sexual affinities and, 84–86; during rut, 28, 89–90; sexually motivated, 84, 86; as sexual segregation basis, 22, 91; social, 84–86; territorial, 6–7; toward predators, 39, 42; via dominance, 28

Akaike's information criteria (AIC), 63–64

alarm and alert behaviors, 34, 37–39, 40, 42, 60

altruistic behavior, 76, 78, 82

amphibians, 12, 13–14, 23

anisogamy, 104

antelopes. *See* African antelopes

anthropogenic disturbances: on birth sites, 124; as habitat fragmentation cause, 124–26; migration effects, 124

antipredator behaviors: in absence of predators, 75, 99, 102; energetic costs, 88; of females, 39–40, 66, 68–69, 99–100; sex differences in, 33, 34, 35–36; sexual segregation as, 75, 76

antipredator tactics, mother-young relationship and, 39–40

antlers: age and body mass correlation, 26, 27; casting/shedding, 76, 84; habitat relationship, 32; protection of, 1, 76, 78, 79–80, 82–83; as secondary sexual characteristic, 2; size, 113

artiodactyls (even-toed ungulates), 2; density dependence, 5; digestive system, 19; mouth morphology, 94; relation to cetaceans, 12; sexual dimorphism, 21, 23, 24–26; sexual segregation, 16–18, 19, 21. *See also specific genera and species*

Asia: deforestation in, 11; ungulate distribution in, 2

asses, 2, 21

Australia, ungulate introduction into, 2

avoidance behavior, 26, 84, 86, 92; intersexual, 26, 86, 90, 94, 98, 119; of intersexual competition, 26, 86, 90, 94, 98; toward predators, 68–69, 88. *See also* spatial segregation

Bateman's Principles, 104, 105

bats (Chiroptera), 12, 15, 19, 23, 80; little brown (*Myotis lucifugus*), 80

bears: black (*Ursus americanus*), 82; brown (*Ursus arctos*), 82; grizzly (*Ursus arctos*), 6, 36, 82

behavioral processes, in sexual segregation, 86

Bell-Jarman Principle, 29–30, 31, 41, 76–77, 83

bighorn sheep (*Ovis canadensis*): anthropogenic disturbances, 124–25; antipredator tactics, 36; birthing period, 48; birth sites, 124; diurnal activity budget, 86; endangered subspecies, 6, 70–71; forage–predation risk trade-offs in, 102; group size as predator defense, 35–36; landscape-based sexual segregation, 50–51; male reproductive success, 27; parturition synchronization, 100;

morphology, dental and mouth, 2–3, 28–29, 41, 94–95, 102. *See also* digestion / digestive system morphology; sexual dimorphism

mortality: additive, 110, 111, 115, 116; compensatory, 110–11, 115–16; delayed density dependence and, 113; density dependence and, 106; of males, 93; overwinter, 106–7; sex differences, 60, 105–6; vehicle collision-related, 124, 126, 127; weather-related, 112, 115; of young, 113

mountain goats (*Oreamnos americanus*), 4, 18, 36

mountain lions (*Puma concolor*), 6, 8, 40, 41

mountain ungulates: antipredator tactics, 36; birthing locations, 36; gregariousness in, 32, 33; tending bonds, 4. *See also* bighorn sheep (*Ovis canadensis*)

mouth morphology. *See* morphology, dental and mouth

movement, anthropogenic barriers to, 124–26, 127. *See also* gait; migration

mule deer (*Odocoileus hemionus*): activity-patterns analysis, 88–90, 92; alert-alarm behaviors, 38–39; concealment cover use, 100; diet, 51–52; foraging behavior, 94; group size–concealment cover relationship, 34, 35; home range, 64–65; migrations, 5, 55, 78, 124; predation effects, 6, 8; scale-dependent landscape metrics, 64–65; sex-affiliated ranges, 64; sexual aggregation, 49; sexual segregation, 64, 67–68; social group size, 44–45, 46; summer nutrition, 107; tending bonds, 4; terrain-sociality relationship, 32, 33. *See also* black-tailed deer (*Odocoileus hemionus*)

multiple causation, of sexual segregation, 100–102

multi-response permutation procedures (MRPPs), 50, 52, 55, 63, 66, 72

muntjacs (*Muntiacus* spp.), 2, 76

musk deer (*Moschus moschiferus*), 2, 3

muskoxen (*Ovibos moschatus*), 4, 18, 39, 65

natural selection, 7, 21, 75; differentiated from sexual selection, 22; habitat use and, 64; mating system shifts and, 4; for reproductive success, 78; in sexual dimorphism, 21, 22–23; as sexual segregation mechanism, 22, 75, 82; for speed and endurance, 39; for understanding nature, 131

New Zealand, ungulate introduction into, 2

niche-based sexual segregation, 68, 69, 71

niche partitioning, 26, 41, 94, 96–97; Bell-Jarman Principle, 29–30, 31, 41, 76–77, 83; population density and, 56, 58; space, habitat, and diet axes in, 98–99, 103

niche-partitioning hypothesis, 98–99, 100–101, 103

niches, 22; dietary, 53–54

nitrogen: fecal content analysis, 54–55, 57–58; forage content, 118

normalized difference vegetation index (NDVI), 55, 58

North America, ungulates in: conservation initiatives, 9–10, 20; deforestation and, 11; distribution, 2; historic declines, 9–10, 20

nutrient cycling, 7–8, 20

nutritional condition: assessment framework, 112–13; female-female competition and, 106; during parturition, 55

nutritional requirements: sexual dimorphism–body size and, 77; tradeoff with predation risk, 101–2

nyala (*Tragelaphus angasii*), 77

observational studies, 47, 81, 87, 130

okapi (*Okapia johnstoni*), 2, 3

omnivores, 2

ontogenetic (developmental) hypothesis, 73–76, 82

open-land species: antipredator behaviors, 33–34, 39–40, 42; gregariousness in, 25, 32, 40; predators' hunting styles, 41; sexual size dimorphism evolution, 25

oryx (*Oryx* spp.), 24

overgrazing, 8, 108, 113

pair territories, 3

parasites, 12, 77–78, 113

parental investment, 104–5, 115, 116, 118, 134

parks, for conservation, 20

parturition: antipredator tactics during, 36, 37; diet, 69–70; habitat selection for, 65–66; nutritional/energy requirements, 101, 107; population density and, 55; predation risk and, 99, 100; seasonality, 45; sexual segregation and, 48, 49, 52, 57, 96; spatial segregation during, 69; synchronization, 100, 106

patch richness, 64–65

patch size, 64–65; coefficient of variation, 64–65

paternal investment, 4–5, 19, 25, 104–5, 134; population dynamics and, 105–16; spatial segregation and, 105

peccaries (Tayassuiidae), 2, 34; collared (*Pecari tajacu*), 24; white-lipped (*Tayassu pecari*), 34

pedicles, 2

pelage markings, use in alarm behavior, 37–38

Perissodactyla (odd-toed ungulates), 2–3, 10–11, 19, 24, 74, 80

pheromones, 37

philopatry, 65

phylogenetic inertia, 21

physical condition assessment, 108

physiology, relation to sexual segregation, 22

pigs, 2

pikas (Lagomorpha), 16, 19

pinnipeds: sexual dimorphism, 23, 26. *See also* sea lions; seals; walrus

plant communities: effects of herbivory on, 7–8, 20, 78; productivity, 32